Series Editor

Prof. Dr. Michael J. Parnham
PLIVA
Research Institute
Prilaz baruna Filipovica 25
10000 Zagreb
Croatia

Fatty Acids and Inflammatory Skin Diseases

Jens-Michael Schröder

Editor

Springer Basel AG

Editors

Prof. Dr. Jens-Michael Schröder
Department of Dermatology
University of Kiel
Schittenhelmstr. 7
D-24105 Kiel
Germany

A CIP catalogue record for this book is available from the Library of Congress, Washington D.C., USA

Deutsche Bibliothek Cataloging-in-Publication Data
Fatty acids and inflammatory skin diseases / ed. by J.-M. Schröder · Springer Basel AG ,1999

(Progress in inflammation research)
ISBN 978-3-0348-9762-4   ISBN 978-3-0348-8761-8 (eBook)
DOI 10.1007/978-3-0348-8761-8

© 1999 Springer Basel AG
Originally published by Birkhäuser Verlag in 1999
Softcover reprint of the hardcover 1st edition 1999
Printed on acid-free paper produced from chlorine-free pulp. TCF ∞
Cover design: Markus Etterich, Basel

ISBN 978-3-0348-9762-4

9 8 7 6 5 4 3 2 1

Contents

Contents

List of contributors

Rosa Antón, Laboratory of Inflammation Mediators, Institute of Research of the Santa Creu and Sant Pau Hospital, S. Antonio Mª Claret 167, E-08025 Barcelona, Spain

Mercedes Camacho, Laboratory of Inflammation Mediators, Institute of Research of the Santa Creu and Sant Pau Hospital, S. Antonio Mª Claret 167, E-08025 Barcelona, Spain

William R. Dunham, Biophysics Research Division, University of Michigan Medical School, Ann Arbor, MI 48109-1055, USA; e-mail: WRDunham@umich.edu

Karsten Fogh, Department of Dermatology, Marselisborg Hospital, University of Aarhus, DK-8000 Aarhus C, Denmark; e-mail: KFOGH@DADLNET.DK

Lars Iversen, Department of Dermatology, Marselisborg Hospital, University of Aarhus, DK-8000 Aarhus C, Denmark

Knud Kragballe, Department of Dermatology, Marselisborg Hospital, University of Aarhus, DK-8000 Aarhus C, Denmark

Anthony I. Mallet, St John's Institute of Dermatology, UMDS, University of London, St.Thomas' Hospital, Lambeth Palace Road, London SE1 7EH, England; e-mail: a.mallet@umds.ac.uk

Cynthia L. Marcelo, Department of Surgery, University of Michigan Medical School, Kresge I, R-5659, Ann Arbor, MI 48109-0592, USA; e-mail: CMARCELO@umich.edu

Hans F. Merk, Hautklinik der Medizinischen Fakultät der RWTH Aachen, Pauwelsstr. 30, D-52074 Aachen, Germany; e-mail: Hans.Merk@post.rwth-aachen.de

Ehrhardt Proksch, Department of Dermatology, University of Kiel, Schittenhelmstr. 7, D-24105 Kiel, Germany, e-mail: eproksch@dermatology.uni-kiel.de

Kerstin Schmidt, Department of Dermatology, Medical Centre Minden, Portastrasse 7–9, D-32423 Minden, Germany

Rudolf Stadler, Department of Dermatology, Medical Centre Minden, Portastrasse 7–9, D-32423 Minden, Germany

Luis Vila, Laboratory of Inflammation Mediators, Institute of Research of the Santa Creu and Sant Pau Hospital, S. Antonio Mª Claret 167, E-08025 Barcelona, Spain

Vincent A. Ziboh, Department of Dermatology, TB 192, School of Medicine, University of California, Davis, CA 95616, USA

Preface

Inflammatory disorders of the skin represent the most common diseases in the general population. The mechanisms of their induction, persistence and possible pharmacological control are still a matter of current investigation. Among the molecule families involved in cutaneous inflammation, lipids, in particular fatty acids and its derivatives, play an important role.

Fatty acids represent major constituents of the skin lipid layer in maintaining barrier function, which is disturbed under inflammatory conditions. We know from studies of deficiencies in nutrition and special diets that unsaturated fatty acid forms are of particular importance. Diets containing unsaturated fatty acids can improve healing and have been used for many years as supplements to treat patients with skin inflammation.

Our knowledge about the molecular mechanisms of involvement of dietary fatty acids in healthy skin physiology and pathophysiology has increased substantially in the last 15 years. It became evident that not only the physical behavior of the outer surfaces, such as membrane fluidity, is strongly influenced by its fatty acid composition. In addition, some unsaturated fatty acids, in particular arachidonic acid, represent the basis of an enormous number of proinflammatory lipid mediators including prostaglandins and leukotrienes. Furthermore, several lipid mediators also act as antiinflammatory compounds.

Skin cells have the capacity to produce a unique and rather characteristic pattern of proinflammatory and antiinflammatory fatty acid derivatives – among them some very recently discovered mediators – which are strongly regulated by different enzymes that control their production and metabolism. Proinflammatory fatty acid metabolites have long been proven to play a role in two major skin diseases, psoriasis and atopic dermatitis. This led to the development of drugs that interfere with the synthesis of these mediators or inhibit their biological functions, and which have been investigated for clinical use at least in part.

This volume is intended to summarize our knowledge about the role of fatty acids and their derivatives in skin physiology and skin pathophysiology, with a particular focus on proinflammatory fatty acid derivatives. New concepts outlined in this volume may stimulate research in fatty acid synthesis regulation, research that modern molecular biological approaches and modern fatty acid detection technology make possible on a high scientific level. Findings may show the way to novel future strategies for pharmacologic intervention in inflammatory skin diseases.

Jens-Michael Schröder

Biosynthesis of fatty acids in the skin and their role in epidermal barrier function

Ehrhardt Proksch

Department of Dermatology, University of Kiel, Schittenhelmstr. 7, D-24105 Kiel, Germany

Introduction

Fatty acids and their derivatives are important cellular structural components and a form of energy storage [1]. In the skin extracellular fatty acids are also part of the intercellular stratum corneum lipid layers. These multilamellar lipid bilayers regulate the permeability barrier that prevents excessive water loss and the entry of harmful substances into the skin [2]. Therefore, the epidermal keratinocyte is highly active in synthesis of lipids, including free fatty acids. The synthesis in the skin responds to various pathological conditions, including disturbed permeability barrier function [3]. However, the essential fatty acids, linoleic acid and arachidonic acid, must be aquired from the circulation [4, 5].

Fatty acid structures

Fatty acids are carboxylic acids with long hydrocarbon tails. Hundreds of different kinds of fatty acid exist [6]. Skin contains free fatty acids as well as fatty acids bound in triglycerides, phospholipids, glycosylceramides and ceramides (sphingolipids). Whereas the chain length of free fatty acids in all epidermal layers ranges from C_{12} to C_{24}, with C_{16} to C_{18} representing the major species in the epidermis, the absolute amount varies and depends on the epidermal layer or on the mammalian body region [4, 7]. Most of the phospholipid derived fatty acids present in the nucleated cell layers of the epidermis are also in the C_{12} to C_{24} range excluding $C_{16:1}$ (C_{16} chain length, monounsaturated). During differentiation of keratinocytes, $C_{16:1}$ and $C_{18:1}$ decreases and $C_{18:2}$ increases. Phospholipids and fatty acids are major components of the cell membrane. In the stratum corneum these lipids are located in the intercellular (extracellular) lipid bilayers. The small amount of phospholipids ($< 5\%$) remaining in the stratum corneum contains a large proportion of long-chain fatty acids [7].

Mechanism of fatty acid synthesis

The *de novo* synthesis of fatty acids is relatively low in most cells, because cells tend to rely primarily on fatty acids incorporated from the plasma [1]. Fatty acids are synthesized by successive condensations of two-carbon units. Acetyl coenzyme A (CoA) is carboxylated to a three-carbon fragment, malonyl-CoA, by the action of acetyl-CoA carboxylase. This is the rate-limiting enzyme in fatty acid synthesis. Malonyl-CoA is the immediate substrate for addition to the growing fatty acid backbone. Of the three carbons in malonyl-CoA, with each condensation cycle, two carbons are added to the fatty acid being synthesized and the third carbon is lost as CO_2 (Fig. 1) [8, 9]. The fatty acid synthesis of animals, once thought to be a large aggregate of individual enzymes in multienzyme complexes, is now known to be two polypeptides with many different catalytic functions [1, 10].

Acetyl-CoA carboxylase is generally regarded as the rate-limiting and controlled enzyme in the biosynthesis of long-chain fatty acids, and both the amount of the enzyme and its catalytic activity are regulated [8]. Short-term regulation, which takes place within minutes, has been attributed to both an allosteric mechanism and to covalent modification in the form of phosphorylation and dephosphorylation of the enzyme. Acetyl-CoA carboxylase polymerizes to an active filamentous form in the presence of the activator, citrate, and disaggregates in the presence of inhibitory long-chain acyl-CoA molecules. On this basis it was proposed that changes in the concentration of citrate or acyl-CoA may regulate the activity of carboxylase *in vivo* [8, 11]. Further studies suggested that the reversible phosphorylation, causing alterations in the citrate requirement for activation, may be the chief determinant for short-term regulation [8, 9].

The regulation of acetyl-CoA carboxylase has been studied on the gene level [8]. There is a single copy of the gene for acetyl-CoA carboxylase per haploid chromosome set. The gene contains two promoters whose primary transcripts are differentially spliced, resulting in multiple forms of acetyl-CoA carboxylase messenger RNA (mRNA). These mRNA species are different in the 5'-untranslated region but contain the same coding region. Generation of different forms of the mRNA is tissue-specific (e.g. liver, fat tissue, mammary gland) and is controlled by physiological conditions. Two promoters contain an extensive array of cis elements that perceive changes in the cellular environment signaling repression and induction of long-chain fatty acid synthesis. The gene responds to various lipogenic signals (e.g. starvation and refeding or pregnancy and lactation), and the same coding sequence is present in all acetyl-CoA carboxylase mRNA species [8, 9, 11]. This suggests that the biosynthesis of fatty acids required for multiple functions in the cells is primarily transcriptionally regulated.

In addition to acetyl-CoA carboxylase a second enzyme called fatty acid synthase can be rate limiting for fatty acid synthesis [10]. The fatty acid synthase is a multienzyme complex responsible for the synthesis of palmitic acid from acetyl-CoA

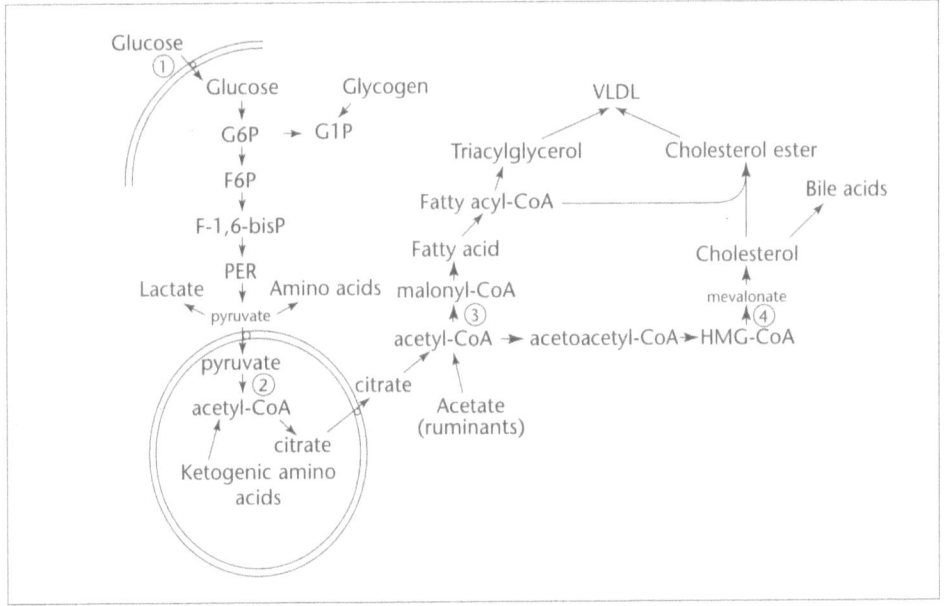

Figure 1
Major pathways of fatty acid and cholesterol synthesis in mammals. (1) glucose transport across the plasma membrane, (2) pyruvate dehydrogenase, (3) acetyl-CoA carboxylase and (4) HMG-CoA reductase.

and malonyl-CoA. The complex consists of two identical proteins of M_r 250,000. Three major functional domains are known. The fatty acid synthase complementary DNA (cDNA) for the chicken liver and the entire rat has been cloned and sequenced [12, 13].

Also, the tissue distribution of human fatty acid synthase mRNA has been studied. Fatty-acid synthase mRNA is ubiquitously distributed in human tissue, with high levels in liver, lung and intraabdominal adipose tissue. The fatty acid synthase gene was localized to human chromosome 17q24-25 and mouse chromosome 11 [14].

Most metabolic reactions of fatty acids in cells involve fatty acids in their activated or fatty acyl-CoA forms [8] (pathway 3, Fig. 1). Mammalian cells contain enzymes known as fatty acyl-CoA synthetases or fatty acid-CoA ligases to convert fatty acids to their corresponding fatty acyl-CoAs. One enzyme, known as nonspecific long-chain acyl-CoA synthetase, is capable of synthesizing a range of fatty acids from 14 carbons to 22 carbons in length. A separate acyl-CoA synthetase, specific for arachidonate and other fatty acids which can be oxygenated to eicosanoids, has been shown to be distinct from the nonspecific long-chain acyl-CoA synthetase. Evidence also exists for a fatty acyl-CoA synthetase specific for long-chain fatty acids greater than 22 carbons in length. The acyl-CoA synthetases are membrane-bound enzymes in the microsomes of cells [8, 11].

Though mammalian cells are able to *de novo* synthesis of fatty acids, they also take up fatty acids from the circulation [1]. The uptake of fatty acids depends on dissociation from serum albumin, transport through the plasma membrane and binding to cytosolic binding proteins. Transport through the plasma membrane by passive diffusion or receptor-mediated uptake has been proposed, though not studied in detail [15].

Fatty acids which enter the cell are preferably esterified into phospholipids [1]. When provided in excess, they are incorporated into newly synthesized triglycerides and stored as lipid droplets in the cells. Very little of the fatty acids in cells is used to form cholesteryl esters, the storage form of cholesterol in the cells [1]. The immediate fatty acid substrate for esterification into phospholipids, triglycerides and cholesteryl esters is fatty acyl-CoA. Acyltransferase enzymes are responsible for the incorporation of fatty acyl-CoA into phospholipids and into triglycerides. A specific enzyme in the cell, acyl-CoA-cholesterol acyltransferase or ACAT, is responsible for the formation of cholesteryl esters from fatty acyl-CoA and cholesterol in cells. An analogous enzyme, lecithin-cholesterol acyltransferase or LCAT, can be found in blood. LCAT catalyzes a reaction in which the fatty acid is removed from a phosphatidylcholine (lecithin) molecule in a lipoprotein and provided to cholesterol to form a cholesteryl ester [1].

Desaturation and elongation of fatty acids

The desaturation of fatty acids in mammalian cells is catalyzed by a series of microsomal enzymes that require molecular oxygen and NADH [1]. There are several distinct desaturase enzymes that insert double bonds at specific positions in fatty acid molecules. Virtually all cells, including those of mammalians, exhibit Δ9 desaturase activity (insertion of a double bond between the ninth and tenth carbons). With this enzyme, cells are capable of synthesizing monounsaturated fatty acids, such as oleate (a Δ9 monounsaturated fatty acid) from the saturated fatty acid stearate [1]. In the desaturation of fatty acids in mammalian cells, fatty acyl-CoA substrates that already contain double bonds can only accommodate a new double bond between the last double bond and the carboxyl group but not between the last double bond and the methyl end of the molecule [1, 16]. This indicates that unsaturated fatty acids cannot change ω or Δ families, as the family designation is determined by the number of carbons between the methyl end of the molecule and the nearest double bond [1].

Fatty acyl-CoA molecules can be elongated by incorporating two carbon units. There are at least two different microsomal elongation enzymes, one for saturated and another for unsaturated fatty acids. Mammalian cells are able to elongate 16 carbon fatty acids to 18 carbon fatty acids and to elongate 18 carbons fatty acids to corresponding 20 carbon fatty acids [1]. An important desaturation/elongation/ desaturation cascade begins with ω-6 fatty acid linoleate as the primary substrate. Linoleate (18:2 Δ9,12) can be desaturated by the insertion of a double bond between the sixth and seventh carbon atom to produce γ-linoleate (18:3, Δ6,9,12) by a Δ6 desaturase. A deficiency in Δ6-desaturase activity has been postulated to be important for the pathogenesis of atopic dermatitis [25, 26]. γ-Linoleate is converted by the action of elongase, with the addition of two carbons, to a fatty acid with 20 carbons and three double bonds to result in dihomo-γ-linoleate (20:3 Δ8,11,14). Finally, 20:3 can be transformed by the addition of a double bond between the fifth and sixth carbons to arachidonate (20:4 Δ5,8,11,14), [1, 17]. Arachidonate is particularly important because it is the primary fatty acid substrate for the cyclooxygenase and lipoxygenase enzymes that produce eicosanoids, a diverse family of biologically potent compounds. There is very little arachidonate in the diet, and because most of the dietary fatty acids of the ω-6 family are linoleate, an intact pathway between linoleate and arachidonate is essential for the production of eicosanoids.

The arachidonate used for eicosanoid synthesis is localized in specific phospholipids residing in specific membranes of the cells. Arachidonate esterified in phospholipids within nuclear membranes is most readily available for eicosanoid production, but because this pool of arachidonate is small, the bulk of arachidonate for eicosanoid production derives from arachidonate esterified in the phospholipids of the endoplasmic reticulum [1].

Fatty acid synthesis in the skin

The skin is quantitatively an important site of *de novo* lipid synthesis [18]. Using radioactive labelling in intact animals, it was shown that fatty acid synthesis in the skin of rodents makes an important contribution to total body fatty acid synthesis. On a weight basis the epidermal keratinocyte is very active in fatty acid synthesis. Though fatty acid synthesis is quantitatively greater in the dermis in comparison to the epidermis on a total tissue basis, it is likely that a major portion of the dermal synthesis is localized to the pilosebaceous epithelium. Because the epidermal layers comprise only a small fraction of total skin mass (approximately 10% or less), on a weight basis the viable epidermis can be considered among the most active sites of lipid synthesis in the entire body [3, 18, 19].

Within the epidermis, all viable cell layers (stratum basale, stratum spinosum and stratum granulosum) synthesize fatty acids, but the stratum corneum displays very little synthetic capability. In neonatal mice the biosynthetic activity related to a protein basis was higher in the stratum granulosum than in the basal and spinous layers [3, 19].

Systemic regulation of epidermal lipid synthesis

A number of lines of evidence have demonstrated the accumulation of dietary and circulating lipids in the skin [20]. The epidermis contains a high concentration of essential fatty acids that are required for normal barrier function. Deficiency of essential fatty acids results in both perturbed barrier function and increased DNA synthesis leading to epidermal hyperplasia [5, 21–24]. These essential fatty acids can only be derived from dietary sources because mammalian cells lack the ability to synthesize these lipids *de novo* [20]. Similarly, ingested ω-3 polyunsaturated fatty acids (fish oils), which cannot be synthesized by mammals, also accumulate in the esterified fatty acid fraction found in the epidermis. Finally, the epidermis lacks the enzyme necessary to convert linoleic acid into arachidonic acid (Δ^6-desaturase) and hence has to take up this fatty acid from the circulation [25, 26]. It is known that dietary and circulating fatty acids can be delivered to the skin, but neither the quantitative impact of this delivery nor the effect of these fatty acids on cutaneous fatty acid metabolism is known except in gross deficiency states such as essential fatty acid deficiency [3]. The effect of dietary fatty acids or triglycerides on epidermal lipid synthesis has not been extensively studied. However, adding glucose to whole skin slices or epidermal samples *in vitro* increased fatty acid synthesis. In contrast, starvation also inhibits lipid synthesis. The mechanism accounting for this glucose stimulation of lipid synthesis is uncertain, but it may be that the metabolism of glucose provides the substrate needed for fatty acid synthesis (acetyl-CoA) or necessary cofactors (NADPH derived from the pentose phosphate shunt pathway) [1].

Hormones can affect skin lipid synthesis. This has been shown for thyroid hormones, androgens, estrogens and corticosteroids. In hypothyroid rats the synthesis of phospholipids, sterol esters, free fatty acids, triglycerides and free sterols is decreased in the epidermis [3, 18]. Similary, the synthesis of lipids in human keratinocytes in culture is also decreased when cells are grown in a medium that does not contain thyroid hormone. It has been shown that thyroid hormones and corticosteroids exert their effects in fatty acid synthesis on the level of the enzyme fatty acid synthase [27]. Though these data indicate that epidermal lipid synthesis may be influenced by systemic factors, it is also apparent that the physiological regulation of epidermal lipid synthesis is unlikely to be mediated primarily by systemic events. Systemic factors affect epidermal lipid synthesis most dramatically when there are major perturbations (hyperthyroidism, diabetes or starvation) [3].

Keratinocytes require the essential fatty acid linoleic acid for the synthesis of stratum corneum membrane lipids [21]. A plasma membrane fatty acid-binding protein is postulated to mediate cellular fatty acid uptake in hepatocytes and several other tissues [15]. A similar mechanism has also been found in keratinocytes. Keratinocytes differ from both hepatocytes and fibroblasts by more rapid uptake of linoleic acid in comparison to oleic acid [15, 28].

The permeability barrier of the skin

In skin fatty acids represent not only important components of cell membranes but also major parts of the permeability barrier [2]. The major function of the epidermis is to form a permeability barrier, which prevents excessive loss of the body fluids required for terrestrial life. This barrier is located mainly in the stratum corneum, which is organized into a heterogenous two-compartment system of protein-enriched cells embedded in lipid-enriched intercellular (extracellular) membrane bilayers. The lipid synthesis required for barrier function occurs in the keratinocytes in all nucleated layers of the epidermis [19]. Lipids are stored and delivered by epidermal lamellar bodies (Odland bodies, keratinosomes). Lamellar body formation is first visible ultrastructurally at the level of the spinous layers. In the upper granular layer the contents of the lamellar bodies are secreted into the intercellular domains of the stratum corneum [21]. The lamellar bodies in the living epidermal layers mainly contain cholesterol, phospholipids and glucosylceramides [29]. Hydrolytic enzymes, which are also located in the lamellar bodies, convert phospholipids and glucosylceramides to free fatty acids and ceramides after secretion [30]. Lipids in the stratum corneum contain approximately equal quantities of ceramides, cholesterol and free fatty acids, as well as lesser amounts of nonpolar lipids and cholesterol sulfate. This mixture forms the membrane bilayer system, which regulates barrier function [21] (Fig. 2).

As mentioned above, there are free fatty acids as well as fatty acids bound to triglycerides, phospholipids, ceramides and glucosylceramides in the skin. The free fatty acid content is only 7% in the basal and spinous layers, increases to 9% in the granular layers and then to 26% in the stratum corneum. In parallel, triglycerides and phospholipids, which generate fatty acids by hydrolysis, decrease from 12 to 24% in the nucleated layers, respectively, to small amounts in the stratum corneum (phospholipids 6%) [2, 29].

Of special importance for the permeability barrier of the skin are sphingolipids (ceramides) [31]. Sphingolipids are prominent components of cellular membranes, lipoproteins and other lipid-rich structures [3]. In the epidermis their common backbones are the long-chain components sphingosine (trans-4-sphingenine) and phytosphingosine (4-D-hydroxysphinganine), with lesser amounts of sphinganine (dihydrosphingosine) and other homologs of these compounds [32–34]. In contrast, keratinocytes in rodents contain only sphingosine as the long-chain base [3, 5]. Epidermal sphingolipids represent 7,3% of total lipid in the basal layers, increasing to about 15% in the stratum granulosum, 30% in the lower stratum corneum and reaching 40% in the outer stratum corneum [29]. Thus, the transformation of the stratum granulosum into the stratum corneum is accompanied not only by a depletion of phospholipids but also by an increase in total sphingolipids. Though glycosphingolipids are present in small quantities in the stratum basale and stratum granulosum, where they are localized to the lamellar bodies, they are virtually absent in the stratum corneum [2, 29].

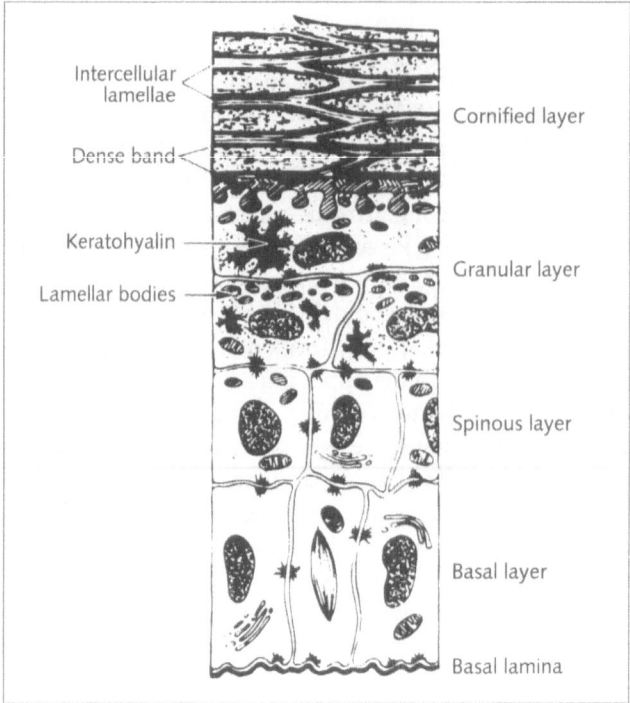

Figure 2
Cross-section of epidermis. Lipids are synthesized in the nucleated layers, assembled in the lamellar bodies and extruded into the extracellular spaces during transition from stratum granulosum to stratum corneum. Stratum corneum consists of corneocytes and intercellular lipids.

Both glycosphingolipids and ceramides are amphipathic molecules, able to fulfill the structural role of phospholipid in maintaining a stable membrane phase in the intercellular spaces of stratum corneum. Long-chain saturated fatty acids (> C_{20}) esterified to glycosphingolipids and ceramides have high melting points and are stable to oxidation, and therefore withstand wide ranges of temperature, ultraviolet (UV) irradiation and oxidation processes [32, 34]. Ceramides are of special interest because this structurally heterogeneous group of substances represents the major polar lipid from which the extracellular membrane structures of the stratum corneum are presumably constructed [2, 5]. Epidermal ceramides represent a unique, heterogenous group of lipids. A series of seven ceramides and an equivalent series of glucosylceramides have been characterized from pig and human epidermis [31–34]. These are composed of the long-chain base sphingosine, with lesser amounts of phytosphingosine and an amide-linked nonhydroxy or α-hydroxy fatty

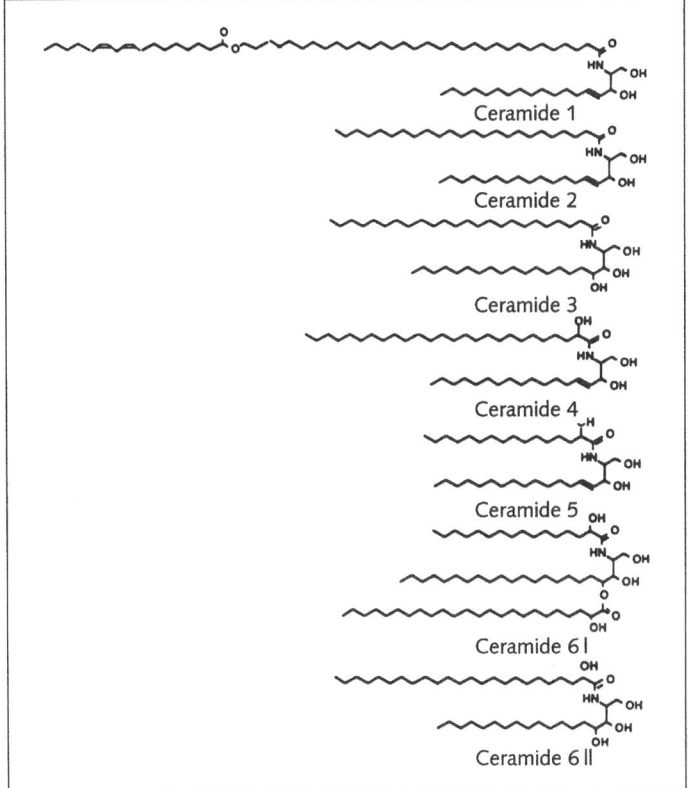

Ceramide 1

Ceramide 2

Ceramide 3

Ceramide 4

Ceramide 5

Ceramide 6 I

Ceramide 6 II

Figure 3
Structure of human stratum corneum ceramides containing different fatty acids.

acid (about 50% α-hydroxylated) [32, 34]. In the stratum granulosum and stratum corneum, long-chain fully saturated fatty acids (C_{22}–C_{26}) predominate in the sphingolipid fraction, whereas other esterified neutral lipids or free fatty acids display a more typical spectrum of fatty acids, with a predominance of $C_{18:1}$ and $C_{16:0}$ [35] (Fig. 3).

One of the most unusual and interesting sphingolipids described to date, and apparently unique to the epidermis, consists of a glycosylated and nonglycosylated sphingosine base with an amide-linked, long-chain, nonhydroxy and α-hydroxy acid, with an additional ω-esterified nonhydroxy acid (primarily $C_{18:2}$) [5, 32]. The amide-linked fatty acid ranges from 29 (in human) to 35 carbon atoms in porcine epidermis, with two hydroxyl groups and two as yet unlocalized double bonds [34]. These species of acylsphingolipids are present in the epidermis of all terrestrial mammalians examined to date, including porcine and human epidermis. Because the predominant ω-linked, esterified fatty acid is linoleic acid, this may indicate a role for these molecules in barrier function. As noted above, the sterified acids of the acyl-sphingolipids contain a high proportion of linoleic acid, ranging from about 35% in mice, to 50% in humans, to 77% in pigs [5, 34].

Local regulation of epidermal lipid synthesis and its role of the barrier

As mentioned above, the stratum corneum permeability barrier is primarily composed of extracellular lipids, mainly cholesterol, sphingolipids and free fatty acids [2]. Thus, one of the factors that might regulate epidermal lipid synthesis would be the status of the permeability barrier. Initial studies by Feingold, Monger and Grubauer in intact mice showed that incorporation of tritiated water into fatty acids and cholesterol was increased after experimental barrier disruption [18, 19, 36]. Barrier disruption induced by topical treatment of the skin with acetone or sodium dodecylsulfate for 3 to 10 min led to an increased epidermal, but not dermal, fatty acid biosynthesis approximately threefold over controls at 1 to 4 h, which returned to normal after 12 h [36]. Since occlusion by a tightly fitting plastic or latex foil immediately performed after barrier disruption blocked the expected increase in *de novo* lipid synthesis in acetone-treated animals, the authors suggested that the water flux (transepidermal water loss, TEWL) provides the molecular signal for *de novo* synthesis [36]. The chronic model of essential fatty acid deficient (EFAD) mice (6–8 weeks on diet) demonstrated a disturbed permeability barrier (increased TEWL) and a twofold increase in fatty acid synthesis in comparison to normals. This effect was neutralized when TEWL was normalized by occlusion, despite the presence of ongoing EFAD [36].

Further studies determined the enzymatic basis for this increase in fatty acid synthesis. Acute barrier disruption by acetone treatment or tape stripping increased rate-limiting enzyme activities, acetyl-CoA carboxylase and fatty acid-synthase activities in the epidermis (+43–69%) [37]. Provision of an artificial barrier by occlusion with an impermeable membrane prevented the increase in acetyl-CoA carboxylase and fatty acid-synthase activities, indicating that the increased activity was dependent on an increase in TEWL and could not be attributed to nonspecific effects [37]. In addition, chronic barrier disruption by EFAD increased acetyl-CoA carboxylase (+127%) and fatty acid-synthase (+49%) activities in the epidermis. Again, occlusion with a water vapor-impermeable membrane decreased both acetyl-CoA carboxylase and fatty acid-synthase activities towards normal [37]. These results indicate that the increase in fatty acid synthesis that occurs in the epidermis after barrier disruption is due to a coordinate regulation in increase in the activities of both epidermal acetyl CoA-carboxylase and fatty acid synthase [37, 38].

To further delineate the role of the key enzymes in fatty acid synthesis for epidermal permeability barrier function, Mao-Qiang et al. applied 5-(tetradecyloxy)-2-

furancarboxylic acid (TOFA), an inhibitor of acetyl-CoA carboxylase, after disruption of the barrier by acetone or tape stripping [38]. TOFA inhibits epidermal fatty acid synthesis by 50% and significantly delays barrier recovery. Moreover, coadministration of palmitate with TOFA normalizes barrier recovery, indicating that the delay is due to a deficiency in bulk fatty acids [38]. Furthermore, TOFA treatment also delays the return of lipids to the stratum corneum and results in abnormalities in the structure of lamellar bodies, the organelle which delivers lipid to the stratum corneum. In addition, the organization of secreted lamellar body material into lamellar bilayers and stratum corneum membrane structures is corrected by coapplication of palmitate with TOFA. These results demonstrate a requirement for bulk fatty acids in barrier homeostasis [37, 38].

Harris et al. examined the regulation of acetyl-CoA carboxylase and fatty acid synthase on the mRNA level in the epidermis by Northern blotting after barrier disruption by acetone or tape stripping. They found a threefold increase in the mRNA level for both these enzymes, which was prevented by occlusion with an impermeable membrane [39]. The exact molecular mechanism by which the epidermal permeability barrier requirements regulate the expression of these genes involved in lipid synthesis is known only in part.

The signal transduction mechanisms for the regulation of fatty acid synthesis in the skin and the recovery of permeability barrier function have been related to ion concentration in the epidermis and to cytokines: a gradient in the concentration of calcium ions has been described in the epidermis, with the highest concentration of calcium ions in the nucleated layers of the outer epidermis and the lowest in the basal layer [40, 41]. Following barrier disruption, the calcium gradient is abolished, returning in parallel with the formation of a functional barrier. Increasing the external concentration of calcium and potassium ions following barrier disruption reduces the return of lipid to the stratum corneum and inhibits barrier recovery [42]. But it remains unknown whether calcium ions influence the expression of lipid synthesizing enzymes, genes and/or the activation of transcription factors.

Several studies on the role of cytokines in permeability barrier repair have been performed. These studies showed that artificial barrier disruption induces or increases the expression of several cytokines in the skin at the mRNA and protein levels. Pronounced effects have been noted on interleukin-1α (IL-1α) and tumor necrosis factor (TNF) expression after barrier disruption [43, 44]. We have shown that application of TNF or IL-1α after experimental barrier disruption leads to enhanced permeability repair (E. Proksch et al., unpublished observations). It was also shown that systemic administration of TNF and IL-1α to Syrian hamsters influences systemic cholesterol metabolism [45]. In addition, the induction of nerve growth factor expression is regulated by barrier status in murine epidermis [46]. However, it is presently unknown whether cytokines and growth factors specifically regulate fatty acid synthesis in the skin. Also, the signals for the esterification of fatty acids to form triglycerides, phospholipids or ceramides are unknown.

In summary, fatty acids are an important part of the physical stratum corneum permeability barrier. The stratum corneum contains free fatty acids as well as fatty acids esterified to a sphingolipid backbone. The epidermis is very active in lipid synthesis; only essential fatty acids need to be taken up from circulation. Fatty acid synthesis in the skin is regulated by permeability barrier function. Signals for permeability barrier recovery, including fatty acid synthesis, are calcium ion concentrations as well as cytokines.

References

1   Laposata M (1995) Fatty acids: biochemistry to clinical significance. *Am J Clin Pathol* 104: 172–179

2   Elias PM (1983) Epidermal lipids, barrier function, and desquamation. *J Invest Dermatol* 88: 445–495

3   Feingold KR (1991) The relation and role of epidermal lipid synthesis. In: Elias PM, Havel RJ, Small DM (eds): *Advances in lipid research: Skin lipids*. Academic Press, San Diego, 57–82

4   Schürer NY, Plewig G, Elias PM (1991): Stratum corneum lipid function. *Dermatologica* 183:77–94

5   Wertz PW, Cho ES, Downing DT (1983) Effect of essential fatty acid deficiency on the epidermal sphingolipids of the rat. *Biochem Biophys Acta* 735: 350–355

6   Jie MS (1993): The synthesis of rare and unusual fatty acids. *Prog Lipid Res* 32: 151–194

7   Schürer NY, Elias PM (1991) The biochemistry and function of stratum corneum lipids. In: Elias PM, Havel RJ, Small DM (eds): *Advances in lipid research: Skin lipids*. Academic Press, San Diego, 27–56

8   Hardie GH (1989): Regulation of fatty acid synthesis via phosphorylation of acetyl-CoA carboxylase. *Prog Lipid Res* 28: 117–146

9   Kim KH, Tae HJ (1984) Pattern and regulation of acetyl-CoA carboxylase gene expression. *J Nutr* 124: 1273s–1283s

10 Alberts B, Bray D, Lewis J, Raff M, Roberts K, Watson JD (eds) (1989) *The molecular biology of the cell*, 2nd edn. Garland Publishing, New York, 44–45, 52, 280

11 Jamil H, Madson NB (1987) Phosphorylation state of acetyl-coenzyme A carboxylase. *J Biol Chem* 262: 630–637

12 Holzer KP, Liu W, Hannes GG (1989) Molecular cloning and sequencing of chicken liver fatty and synthase cDNA. *Proc Natl Acad Sci USA* 86: 4387–4391

13 Amy CM, Witkowski A, Naggert J, Williams B, Randhawa Z, Smith S (1989) Molecular cloning and sequencing of cDNAs encoding the entire rat synthase. *Proc Natl Acad Sci USA* 86: 3114–3118

14 Semenkovich CF, Coleman T, Fiedorek FT (1995: Human fatty acid synthase mRNA:

tissue distribution, genetic mapping, and kinetics of decay after glucose deprivation. *J Lipid Res* 36: 1507–1521

15 Schürer NY (1996) *Fettsäure bindende Proteine humaner Keratinozyten.* Hensel-Hohenhausen-Verlag, Egelsbuch (Germany), 25–30

16 Buddecke (1977) *Grundriss der Biochemie,* 5th edn. De Gruyter, Berlin, 203–204

17 Sprecher H, Luthrea DL, Mohammed BS, Baykousheva SV (1995) Reevaluation of the pathways for the biosynthesis of polyunsaturated fatty acids. *J Lipid Res* 36: 2471–2477

18 Feingold KR, Brown BE, Lear SR, Moser AM (1983) Localisation of *de novo* sterologenesis in mammalian skin. *J Invest Dermatol* 81: 365–369

19 Monger DJ, Williams ML, Feingold KR, Brown BE, Elias PM (1988) Localisation of sites of lipid biosynthesis in mammalian epidermis *J Lipid Res* 29: 603 – 612

20 Protey C (1976): Essential fatty acids and the skin. *Br J Dermatol* 94: 579–583

21 Elias PM, Brown BE (1978): The mammalian cutaneous permeability barrier: defective barrier function in essential fatty acid defiency correlates with abnormal intercellular lipid deposition. Lab Invest 39: 574–583

22 Proksch E, Feingold KR, Elias PM (1991): Barrier function regulates epidermal lipid and DNA synthesis. *J Clin Invest* 87: 1668–1673

23 Proksch E, KR Feingold, PM Elias (1992) Epidermal HMG CoA reductase activity in essential fatty acid deficiency: barrier requirements rather than eicosanoid generation regulate cholesterol synthesis. *J Invest Dermatol* 99: 216–220

24. Proksch E, WW Holleran, GK Menon, PM Elias, KR Feingold (1993) Barrier function regulates epidermal lipid and DNA synthesis. *Br J Dermatol* 128: 473–482

25 Horrobin DF, Morse PF (1985): Evening primrose oil and atopic dermatitis. *Lancet* 345: 260–261

26 Horrobin DF (1989) Essential fatty acids in clinical dermatology. *J Am Acad Dermatol* 20: 1045–1053

27 Sprecher H, Luthria DL, Mohammed BS, Baykousheva SV (1995) Reevaluation of the pathways for the biosynthesis of polyunsaturated fatty acids. *J Lipid Res* 36: 2471–2477

28 Schürer NY, Stremmel W, Grundmann JU, Schliep V, Kleinert H, Bass NM, Williams ML (1994) Evidence for a novel keratinocyte fatty acid uptake mechanism with preference for acid: Comparison of oleic and linoleic acid uptake by cultured human keratinocytes, fibroblasts and a human hepatoma cell line. *Biochim Biophys Acta* 1211: 51–60

29 Lampe MA, Burlingame AL, Whitney J, Williams ML, Brown BE, Roitman E, Elias PM (1983) Human stratum corneum lipids: characterization and regional variations. *J Lipid Res* 24: 120–130

30 Menon GK, Grayson S, Elias PM (1986) Cytochemical and biochemical localisation of lipase and sphingomyelinase activity in mammalian epidermis. *J Invest Dermatol* 86: 591–597

31 Wertz PW, Downing DT (1982) Glucolipids in mammalian epidermis: structure and function in the water barrier. *Science* 217: 1261–1262

32 Kerscher M, Korting HC, Schäfer-Korting (1991) Skin ceramides: structure and function. *Eur J Dermatol* 1: 39–43

33 Wertz PW, Downing DT (1990) Metabolism of linoleic acid in porcine epidermis. *J Lipid Res* 31: 1839–1844

34 Robson KJ, Stewart ME, Michelsen S, Lazo ND, Downing DT (1994) 6-Hydroxy-4-sphingenine in human epidermal ceramides. *J Lipid Res* 35: 2060–2068

35 Bouwstra JA, Gooris GS, Dubbelaar TER, Weerheim AM, Izerman AP, Ponec M (1998) Role of ceramide 1 in the molecular organization of the stratum corneum lipids. *J Lipid Res* 39: 186–196, 1998

36 Grubauer G, Feingold KR, Elias PM (1987) Relationship of epidermal lipogenesis to cutaneous barrier function. *J Lipid Res* 28: 746–752

37 Ottey AK, Ladonna C, Grunfeld C, Elias PM, Feingold KR (1995) Cutaneous permeability barrier disruption increases fatty acid synthetic enzyme activity in the epidermis of hairless mice. *J Invest Dermatol* 104: 401–404

38 Mao-Qiang M, Elias PM, Feingold KR (1993) Fatty acids are required for epidermal peremeability barrier function. *J Clin Invest* 92: 791–798

39 Harris IR, Farrell AM, Grundfeld C, Holleran WM, Elias PM, Feingold KR (1997) Permeability barrier disruption regulates mRNA levels for key enzymes of cholesterol, fatty acid and ceramide synthesis in the epidermis. *J Invest Dermatol* 109: 783–789

40 Menon GK, Grayson S, Elias PM (1985) Ionic calcium reservoirs in mammalian epidermis: ultrastructural localisation by ion-capture cytochemistry. *J Invest Dermatol* 84: 508–512

41 Menon GK, Elias PM (1991) Ultrastructural localisation of calcium in psoriatic and normal human epidermis. *Arch Dermatol* 127: 57–63

42 Lee SH, Elias PM, Proksch E, Menon GK, Mao-Qiang M, Feingold KR (1992) Calcium and potassium are important regulators of barrier homeostasis in murine epidermis. *J Clin Invest* 89: 530–588

43 Wood LC, Jackson SM, Elias PM, Grunfeld C, Feingold KR (1992) Cutaneous barrier pertubation stimulates cytokine production in the epidermis of mice. *J Clin Invest* 90: 482–487

44 Nickoloff BJ, Naidu Y (1994) Pertubation of epidermal barrier function correlates with initiation of cytokine cascade in human skin. *J Am Acad Dermatol* 30: 535–546

45 Hardardottir J, Moser A, Menon R, Grunfeld C, Feingold KR (1994) Effects of TNF, Il-1 and the combination of both cytokines on cholesterol metabolism in syrian hamster. *Lymphokine Cytokine Res* 13: 161–166

46 Liou A, Elias PM, Grunfeld C, Feingold KR, Wood LC (1997) Amphiregulin and nerve growth factor expression are regulated by barrier status in murine epidermis. *J Invest Dermatol* 108: 73–77

Arachidonic acid metabolism in skin

Lars Iversen and Knud Kragballe

Department of Dermatology, Marselisborg Hospital, University of Aarhus,
DK-8000 Aarhus C, Denmark

Introduction

Arachidonic acid (AA) (eicosa-5,8,11,14-tetraenoic acid, 20:4 ω6) is a polyunsaturated fatty acid with 20 carbon atoms and 4 double bonds. It is either derived from dietary sources or synthesized by desaturation and elongation of linoleic acid (C18:2) (Fig. 1). Because the mammalian organism cannot introduce double bounds in the fatty acid structure closer to the ω end than ω9, the AA precursors linoleic acid (18:2 ω6) and γ-linoleic acid (18:3 ω3) are considered essential [1]. AA cannot be synthesized locally in the human epidermis because both Δ-6-desaturase and Δ-5-desaturase are absent in the epidermis [2]. AA present in the epidermis must therefore either come from dietary sources or be transported to the epidermis from other endogenous sources such as the liver, which is capable of elongation and desaturation. In the cell AA is stored in the membrane fraction, primarily esterified to phospholipids at the second carbon (*sn*-2) of the phospholipid glycerol backbone [3]. It is released from the phospholipids in the cell membrane by the action of phospholipases.

Phospholipases

Esterified AA is not amenable to metabolic transformations. Therefore, AA has to be released from the cell membrane either by phospholipase A_2 (PLA2) or by a combined action of phospholipase C (PLC) and a diglyceride lipase. PLA_2 catalyzes the hydrolysis of the *sn*-2 fatty acyl bond of phospholipids to liberate free fatty acids and either lysophospholipids or lyso-PAF (platelet-activating factor) depending on type of binding in the 1-position of the phospholipid (Fig. 2). The lysophospholipid may play a role in modulating or activating certain cells [4, 5], and lyso-PAF may act as a precursor of PAF.

The PLA_2s are a diverse class of enzymes with regard to function, localization, regulation, mechanism, sequence and structure [6]. Thus, they carry out essentially

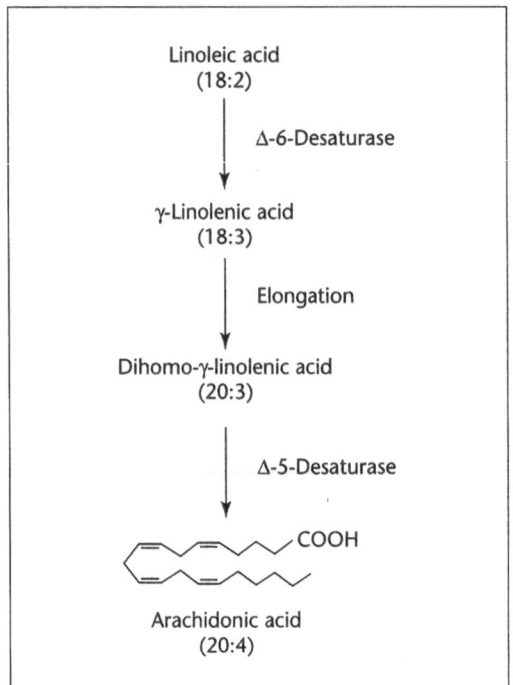

Figure 1
Arachidonic acid synthesis.

the same reaction, namely the hydrolysis of the *sn*-2 fatty acyl bond of phospho-lipids. Traditionally, three different human PLA_2 classes have been described and characterized; pancreatic PLA_2 which is classified as digestive [7], and two regula-tory PLA_2s, secretory (s) PLA_2, [8] and cytoplasmic (c) PLA_2 [9]. Although recent evidence has revealed that the PLA_2s may be a much more diverse group of enzymes than previously believed [10] and that the group designations will undoubtedly need to be adjusted, the designation mentioned above will be used here.

Both the $sPLA_2$ and the $cPLA_2$ are present in human skin [11, 12]. $sPLA_2$s con-sist of a large isozyme family [13]. These 14-kDa isozymes are secretory enzymes and therefore mainly present extracellularly. They are devoid of specificity on *sn*-2 fatty acid and require millimolar concentration of Ca^{2+} for activation [14]. They are believed to be a key link in the inflammatory process, e.g. as an effector system for inflammatory cytokines [15], and both involved and uninvolved psoriatic skin has been shown to contain higher levels of $sPLA_2$ than normal skin [11]. Several cytokines present in the skin have also been reported to modulate the expression of $sPLA_2$. Interleukin (IL)-1β has been seen to induce expression and secretion of $sPLA_2$ in a variety of cells [15], while IL-6 was shown to elevate $sPLA_2$ levels in a human hepatoma cell line [16]. In contrast, transforming growth factor β (TGFβ) inhibits $sPLA_2$ expression in rat mesangial cells [17].

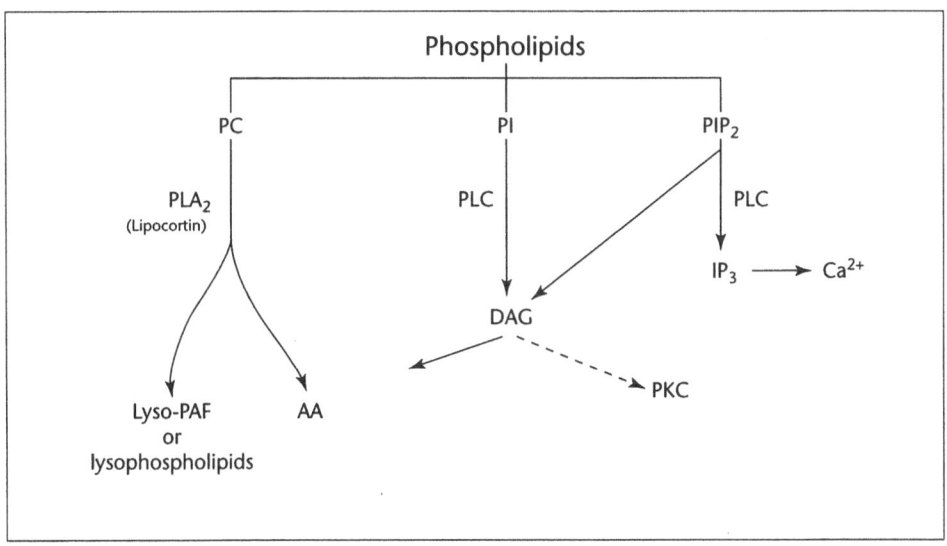

Figure 2
Phospholipase-mediated processes. A dotted arrow indicates modulation of enzyme activity (see text).

cPLA$_2$ has been purified from monocytic cell lines, U937 [18] and RAW264.7 [19], and the complimentary DNA (cDNA) coding for the enzyme was cloned from a U937 cell cDNA library [20]. The molecular weight is 85 kDa as determined by sequencing and cloning [20]. The cPLA$_2$ is translocated from the cytosol to the membrane fraction in a Ca^{2+}-dependent fashion [21] at physiologically relevant Ca^{2+} concentrations. The translocation to the membrane may be a regulatory process augmenting accessibility of the enzyme to the substrate. In contrast to the sPLA2, Ca^{2+} does not play a catalytic role for cPLA$_2$, it seems only to play a role in the association of the enzyme with the membrane phospholipids. An interesting characteristic of cPLA$_2$ with respect to eicosanoid formation is its AA specificity [19, 22], and data have indicated that the cPLA$_2$ is responsible for AA liberation in agonist-stimulated inflammatory cells. The involvement of AA-specific PLA$_2$ in the production of PAF and eicosanoids from 1-O-alkylglycerophosphocholine was suggested because both PAF and eicosanoid biosynthesis were abolished in polymorphonuclear leukocytes [23] and HL-60 cells [24] depleted of AA.

The calcium requirement of PLA$_2$ has led to speculation that PLC activation and the resulting inositol-1,4,5-triphosphate (IP$_3$) and intracellular Ca^{2+} elevation (Fig. 2) precede PLA$_2$ activation. However, several experiments have shown that PLA$_2$ and PLC activation are regulated independently by different G proteins. Agonist-induced AA release was shown to be sensitive to pertussis toxin treatment in mouse 3T3 fibroblasts [25] and a rat thyroid cell line, FRTL-5 [26], whereas phosphatidylinositol-4,5-biphosphate (PIP_2) hydrolysis in the same cells was insensitive to toxin treatment.

In several studies PLA_2 has been suggested to play an important role in inflammatory skin diseases. Intradermal injection of purified sPLA_2 has been shown to induce cellular infiltration, interstitial edema, vascular permeability and hyperemia [27]. Also, in psoriasis PLA_2 has been held responsible at least in part for the elevated AA levels found, and elevated PLA_2 activity has been demonstrated in psoriatic epidermis [28]. Furthermore, a recent report has shown an increased lysophosphatidylcholine content in lesional psoriatic skin [29], supporting the idea of increased PLA_2 activity in inflammatory skin diseases. Investigations on how PLA_2 activity is regulated have also been carried out. In the human epidermis PLA_2 has been suggested to be subject to positive feedback regulation, as its activity was stimulated by prostaglandin E_2 (PGE_2) and prostaglandin $F_{2\alpha}$ ($PGF_{2\alpha}$) [30]. PLA_2 activity may also be regulated by direct phosphorylating activities [31] or indirectly by phosphorylation-dephosphorylation of a soluble 35-kDa PLA_2 inhibitory protein termed lipocortin [32]. It has been suggested that upon phosphorylation, lipocortin loses its inhibitory properties, resulting in expression of PLA_2 activity, and hyperphosphorylation of lipocortin has been suggested as the reason for increased PLA_2 activity in uninvolved psoriatic epidermis [33]. Also, glucocorticosteroids act by inhibition of PLA_2 activity [34] via transcriptional control of lipocortin synthesis [35], and topical glucocorticosteroid application to skin results in reduction of PLA_2 activity [36].

A second pathway for AA release is via breakdown of PIP_2 by the phosphoinositide-specific PLC. Hydrolysis of PIP_2 generates diacylglycerol (DAG) and IP_3 [37] (Fig. 2). DAG can be further degraded by diglyceride lipase to glycerol and free fatty acids, including AA [38] (Fig. 2). Also, PLC activation is a primary event in intracellular signaling [39]. IP_3 stimulates Ca^{2+} mobilization from intracellular stores, and DAG activates protein kinase C (PKC), which orchestrates a cascade of biochemical events that regulate cellular responses such as growth and differentiation.

PLC activity is present in most mammalian cells and tissues including human epidermis [39, 40], and cDNAs coding for five PLC isozymes (α, β, γ_1, γ_2 and δ) have been cloned and sequenced [39]. Epidermal PLC has been demonstrated as a Ca^{2+}-dependent enzyme with maximal activity at pH 7.0. PLC activity in psoriatic plaque has been shown to be increased 188% compared with normal epidermis and may thus contribute to the elevated AA levels observed in this tissue [40].

At least two different mechanisms, phosphorylation of PLC by the tyrosine kinase family and G protein-mediated PLC activation, appear to be involved in regulation of PLC activity [41, 42]. Epidermal growth factor has been demonstrated to activate PLC by augmenting tyrosine kinase activity [41]. The regulatory effect of tyrosine phosphorylation on PLC activity is not yet known. Increased catalytic activity due to tyrosine phosphorylation has been reported [43], but it has also been suggested that phosphorylation increases accessibility of PLC to substrate [44].

Cyclooxygenase

The initial step in AA metabolism by the cyclooxygenase pathway is transformation of AA into prostaglandin H_2 (PGH_2) by cyclooxygenase (COX) or PGH synthase. COX catalyzes a two-step reaction with insertion of two oxygen molecules at C11 and C15, resulting in the formation of PGG_2 [45]. The second step is reductive cleavage at the 15-hydroperoxy group to yield PGH_2 [46]. COX is an iron-containing dimer of 70-kDa subunits localized primarily in the endoplasmic reticulum. COX is inhibited by nonsteroidal antiinflammatory drugs such as aspirin and indomethacin. This inhibition largely accounts for the antiinflammatory and analgesic effects of these agents [47]. Recently, a second isoform of COX was cloned and sequenced [48, 49]. The isoform designated COX-1 is constitutively expressed in cells, whereas the isoform designated COX-2 seems to require specific induction. In general, COX-1 regulates prostaglandin synthesis associated with cellular homeostasis, whereas COX-2 is upregulated in inflammatory conditions and associated with synthesis of proinflammatory prostaglandins [50]. Much attention has therefore been paid to developing specific inhibitors of COX-2. Recently, normal murine epidermis was found to express COX-1 but not COX-2. However, COX-2 could be induced either by acetone treatment or by topical application of the phorbol ester, 12-O-tetradecanoyl-phorbol 13-acetate (TPA) [51]. This is in contrast to normal human epidermis, where COX-2 has been associated with keratinocyte differentiation [52]. In normal human skin, COX-1 immunostaining is observed throughout the epidermis, whereas COX-2 immunostaining increases in the more differentiated, suprabasilar keratinocytes [52]. Furthermore, inducing differentiation of cultured human keratinocytes by raising extracellular calcium leads to increased expression of both COX-2 protein and messenger RNA (mRNA). In contrast, no significant alteration in the expression of COX-1 was seen in response to increased extracellular calcium [52].

PGH_2 has been shown to be a central intermediate in prostaglandin biosynthesis by serving as substrate for several competing enzymes (Fig. 3). PGE synthase or PGH-PGE isomerase catalyzes the rearrangement of the endoperoxide group of PGH_2 to produce PGE2. PGE synthase activity has been localized to the microsomal fraction and shown to require glutathione as a cofactor [53]. As early as 1970, PGE_2 formation in rat epidermis was demonstrated [54], and later rat [55] and human skin [56] were shown to synthesize PGE_2. PGE_2 is the main AA cyclooxygenase product in human epidermis homogenates [57]. PGE_2 formation has also been demonstrated to be important for keratinocyte differentiation. In cultured human keratinocytes which have been induced to differentiate by increased extracellular

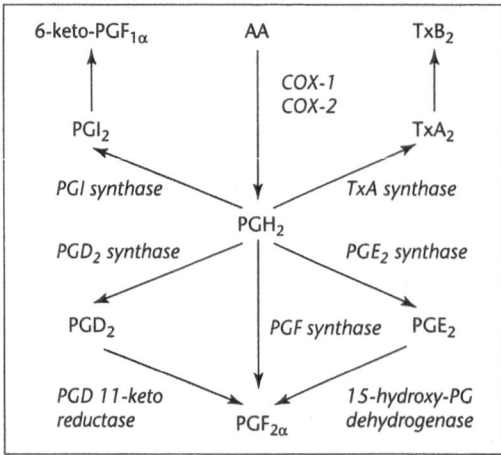

Figure 3
Cyclooxygenase pathway.
Enzymes are indicated in italics.

calcium, increased PGE_2 formation was observed 1 h after the calcium concentration was increased [58].

PGD_2 is formed from PGH_2 either enzymatically or nonenzymatically. Nonenzymatic transformation of PGH_2 into PGD_2 is catalyzed by serum albumin [59]. Also, two specific enzymes, PGD synthases, have been demonstrated. PGD synthases have been localized as cytoplasmic enzymes, unlike most other enzymes involved in prostaglandin formation, which are found predominantly in the microsomal fraction [60]. PGD_2 formation was demonstrated in human skin [57], mouse epidermis, cultured neonatal mouse keratinocytes, rat skin [61] and in guinea pig skin homogenates [62]. Interestingly, PGD synthase has been shown to be present predominantly in the Langerhans' cells in guinea pig and rat epidermis and not in the keratinocytes [61, 63]. Furthermore, PGD synthase was demonstrated in the dermal macrophages and mast cells, and generally mast cells are believed to represent one of the major cellular sources of PGD_2.

There are three possible pathways for the synthesis of $PGF_{2\alpha}$. First, PGF synthase from PGH_2 (Fig. 3) [64]; second, reduction of PGE_2 catalyzed by an NADH-linked 15-hydroxy-PG dehydrogenase [65]; and third, PGD_2 catalyzed by a PGD 11-keto reductase [66]. $PGF_{2\alpha}$ formation has been demonstrated in both rat and human skin [67]. Thromboxane A_2 (TXA_2) is formed from PGH_2 catalyzed by TX synthase. TXA_2 is extremely unstable and therefore rapidly transformed by a nonenzymatic process into TXB_2 [68]. Only very low TXA synthase activity has been found in the skin. Prostacyclin PGI_2, like TXA_2, is an unstable molecule which is rapidly further metabolized into 6-keto-$PGF_{1\alpha}$. The transformation of PGH_2 into PGI_2 is catalyzed by PGI synthase [69]. Very low amounts of $PGF_{1\alpha}$ have been found in studies of AA metabolism in guinea pig skin homogenates, indicating that PGI_2 formation can take place at least in guinea pig skin [62].

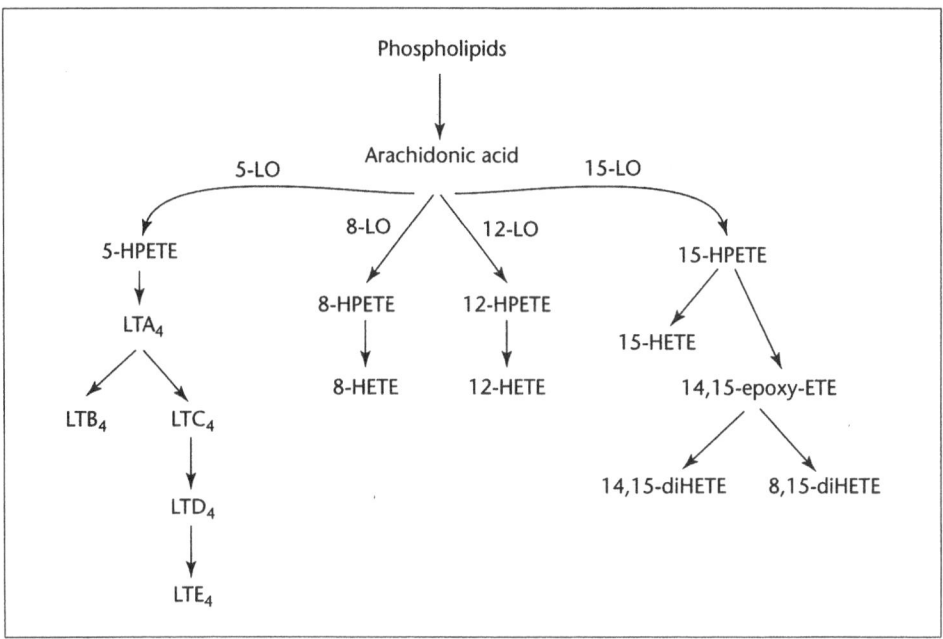

Figure 4

Transformation of arachidonic acid by the 5-, 8-, 12- and 15-lipoxygease pathways.

Lipoxygenases

The initial product of AA made by lipoxygenases (LOs) is the (mono)-hydroperoxy-eicosatetraenoic acid (HPETE) (Fig. 4). 5-,8-,12- and 15-LOs introduce an oxygen molecule into the respective position of the arachidonic acid backbone, giving rise to unstable HPETEs. An HPETE undergoes reduction to its corresponding hydroxy-eicosatetraenoic acid (HETE) and under certain circumstances further oxidation to diHETEs. Among the diHETEs of the 5-LO pathway are the biologically active leukotrienes (LTs).

5-Lipoxygenase

Figure 5 shows the metabolism of AA via the 5-LO pathway. Once AA is liberated from the phospholipids, 5-LO is activated in the presence of adenosine triphosphate (ATP) and Ca^{2+} [70] and translocated from the cytoplasm to the plasma membrane [71] by a Ca^{2+}-dependent mechanism. Activated 5-LO is always membrane-associated [70, 72], and recently a novel membrane-associated 5-LO activating protein

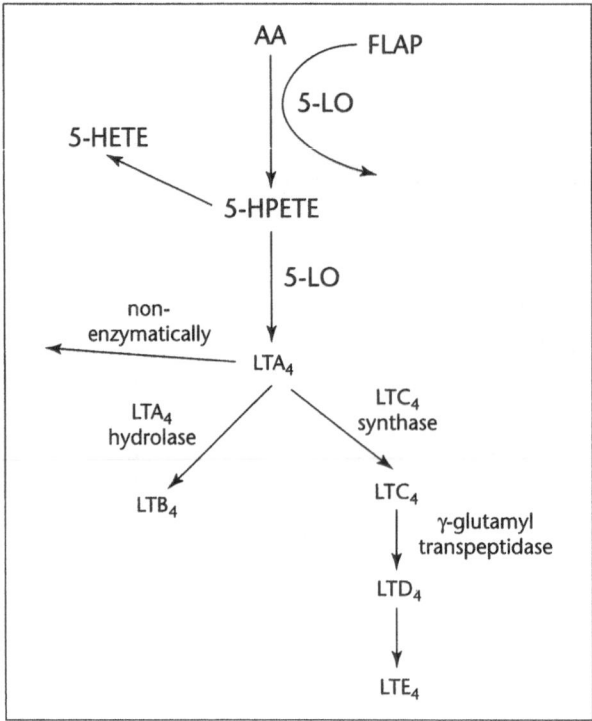

Figure 5
The 5-lipoxygenase pathway.
Arachidonic acid metabolites
are indicated in bold face.

(FLAP) was described [73]. FLAP has been purified from rat neutrophil membranes [73] and cloned from rat basophil leukemia cell and human HL-60 cDNA libraries [74]. It has been identified as an 18-kDa protein [73]. So far all 5-LO-expressing cells investigated have been shown to contain FLAP. Also, transfection experiments have demonstrated that both FLAP and 5-LO must be present in order to transform AA into 5-HPETE [74]. 5-HPETE is then further metabolized by 5-LO into LTA$_4$ [70, 75] or transformed, either enzymatically by a glutathione-dependent peroxidase or nonenzymatically, into 5-HETE [76, 77]. Transformation of AA into LTA$_4$ results in suicide inactivation of 5-LO [71].

The end product of 5-LO activity is LTA$_4$, an unstable allylic intermediate [78, 79] that can be further metabolized both enzymatically and nonenzymatically. Nonenzymatically, LTA$_4$ is metabolized into 5,6-diHETEs and 5,12-diHETEs [80]. The epoxide hydrolase, LTA$_4$ hydrolase, catalyzes the transformation of LTA$_4$ into LTB$_4$ [81], and this step has been shown to be the rate-limiting step in LTB$_4$ formation, at least in rat basophilic leukemia cells (RBL-1) and human neutrophils [82, 83].

LTA$_4$ may also be conjugated with glutathione by LTC4 synthase to yield LTC$_4$ [78]. Successive cleavage by γ-glutamyl transferase [84] and a dipeptidase [85] converts LTC_4 into LTD_4 and LTE_4. Together, LTC_4, LTD_4 and LTE_4 are termed pepti-doleukotrienes.

Although LTs have been ascribed an important role in inflammatory skin diseases like psoriasis and atopic dermatitis [86–89], the capacity of the human skin itself to biosynthesize the LTs has been questioned. It has been reported that freshly isolated human epidermal cells [90, 91] and cultured mouse keratinocytes [92] can synthesize low quantities of LTB_4 as determined by high-pressure liquid chromatography (HPLC) [92] and by radioimmunoassay (RIA) and chemotactic activity [90, 91]. This view was recently supported by Jassen-Timmen et al. [93]. They reported that undifferentiated keratinocytes did not express detectable 5-LO mRNA, 5-LO protein or activity. However, inducing differentiation by shifting the culture conditions resulted in induction of 5-LO gene expression. Thus, induction of 5-LO expression was much more pronounced in keratinocytes derived from the HaCat cell line than in normal human keratinocytes. In contrast, in a study by Brenton et al. [94] measurements of 5-LO protein determined by Western blots or 5-LO mRNA determined by reverse transcriptase polymerase chain reaction analysis in cultured human keratinocytes treated with either IL-1, 1, 25-vitamin D_3, interferon-γ, phorbol 12-myristate 13-acetate (PMA), A23187 or dexamethasone did not demonstrate the presence of 5-LO in human keratinocytes. Furthermore, FLAP was not present in subcellular fractions of these keratinocytes, consistent with the absence of FLAP mRNA in these cells. The lack of 5-LO is in accordance with previous reports from our laboratory [95–97] and others [98] showing no detectable LTB_4 formation in normal human cultured keratinocytes stimulated with A23187 or in freshly isolated human epidermis. It is therefore likely that 5-LO activity in human keratinocytes is very limited.

Transcellular leukotriene synthesis

Although the epidermis cannot form leukotrienes itself from AA, it can contribute significantly to leukotriene formation through transcellular leukotriene synthesis. By this mechanism, LTA_4 formed in one cell type is released and then further metabolized in another cell type (Fig. 6). Human cultured keratinocytes and human epidermis have been shown to transform neutrophil-derived LTA_4 into LTB_4 *in vitro* [95–98]. The key enzyme in transcellular LTB_4 formation in the epidermis is LTA_4 hydrolase (Fig. 6). LTA_4 hydrolase has been localized in human epidermis by activity determination [95–98], inhibition of enzyme activity by bestatin and captopril (known LTA_4 hydrolase inhibitors) [95–97], Western blotting [97] and immunohistochemical staining [99]. Furthermore, LTA_4 hydrolase has been purified and characterized from cultured human keratinocytes and human epidermis [100]. Epidermal LTA_4 hydrolase was characterized as a 70-kDa enzyme with a pI of 5.1 to 5.4 and a pH optimum of 7.5 to 8.5, and it was localized in the cytoplasmic fraction [97]. The amino acid composition has been determined, and all the data obtained from human epidermis and cultured human keratinocytes have been consistent with

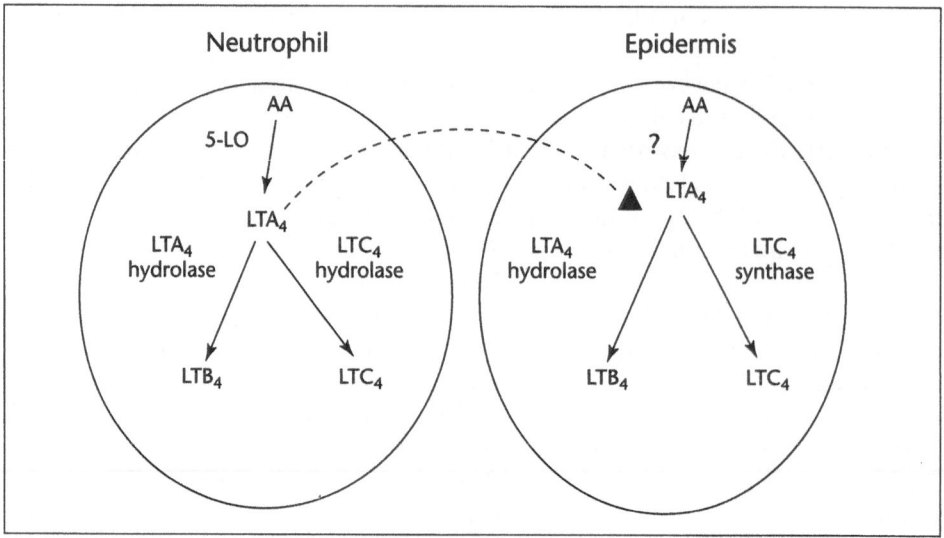

Figure 6
Schematic presentation of transcellular leukotriene synthesis in the epidermis.

data obtained with LTA_4 hydrolase from most other cell types that have been investigated [100]. An interesting characteristic of LTA_4 hydrolase is that it has a dual function, exhibiting both hydrolase activity and peptidase activity. Peptidase activity has been shown against small peptides, including opioids [101, 102]. Another characteristic feature of LTA_4 hydrolase is that it undergoes suicide inactivation when it transforms LTA_4 into LTB_4. This has been demonstrated in several cell types [103], including human epidermis [99], and is due to covalent binding of LTA_4 to the enzyme [104]. Presumably, as a result of suicide inactivation, LTA_4 hydrolase has been shown to be the rate-limiting step in LTB_4 formation, at least in human neutrophils [83] and rat basophilic leukemia cells [82]. Therefore, transcellular LTA_4 metabolism may result in increased LTB_4 formation at an inflammatory site. In inflammatory skin diseases like psoriasis, neutrophils migrate into the epidermis and get into close contact with the keratinocytes. Release of LTA_4 into the extracellular space has previously been demonstrated by activated neutrophils, and very recently it was demonstrated that more than 50% of the LTA_4 formed in neutrophils is released from the cell [105]. Transcellular leukotriene synthesis may therefore be an important mechanism by which the human epidermis can contribute significantly to LTB_4 formation in inflammatory skin diseases.

Similar to transcellular LTB_4 synthesis, transcellular LTC_4 synthesis has been demonstrated as a possible mechanism for LTC_4 formation in the epidermis (Fig. 6)

[96]. LTA$_4$ is transformed into LTC$_4$ by a LTC$_4$ synthase (Fig. 5), which in mouse mastocytoma cells has been shown as a highly specific, membrane bound glutathione-S-transferase (GST) [106]. In the skin a specific LTC$_4$ synthase has, however, never been shown. Several isoforms of GST with activity towards LTA$_4$ have been identified in human and rodent skin [107], and human, rat and mouse skin has been demonstrated to transform LTA$_4$-methyl ester into LTC$_4$-methyl ester [108].

8-Lipoxygenase

8-LO is one of the most recently discovered LOs in the skin. 8-LO activity is best characterized in tissue samples from marine sources [109]. Only brief reports have been published on the enzymatic production of 8-HETE in human tissues and cell types such as human leukocytes [110], human tracheal cells [111] and psoriatic skin [112]. However, 8-LO activity in mouse skin has been determined and characterized [113, 114]. Epidermal 8-LO was shown as a specific 8(S)-LO catalyzing the formation of only 8(S)-HPETE and 8(S)-HETE [113]. Enzyme activity was localized in the cytosolic fraction of the cells in the suprabasal compartment of the epidermis [114] and was stimulated by phosphatidylcholine and lecithin, whereas no requirement of ATP, calcium or NADPH was found [113, 114]. In contrast to the 5-LO, 8-LO is not translocated to the membrane when it is activated. Interestingly, only very low 8-LO activity has been determined in normal mouse epidermis, indicating that the enzyme is not constitutively expressed in the epidermis. However, a prominent dose-dependent induction of 8-LO activity has been demonstrated in mouse epidermis after topical treatment with TPA [114]. These data, together with the fact that 8-HETE is present in psoriatic skin [112], suggest some pathophysiological function of the 8-LO pathway in the skin. Interestingly, it has recently been reported that 8(S)-HETE is a high-affinity ligand for the peroxisome proliferator-activated receptor α (PPARα), whereas 8(R)-HETE was much less potent [115, 116]. PPARs are members of the nuclear receptor superfamily that includes receptors for steroid, thyroid and retinoid hormones. Through dimerization of PPAR with the 9-*cis*-retinoic acid receptor (RXR) and activation of a PPAR response element, transcription of specific genes are regulated. It can be speculated that this is a possible pathway by which 8(S)-HETE exerts some of its possible pathophysiological effects in the skin. It is certainly an area that deserves further investigation in the future.

12-Lipoxygenase

12-HETE has been shown to be one of the main eicosanoids formed by the epidermis [57], and with the discovery of large quantities of 12-HETE in human psoriatic lesions [57, 117], epidermal 12-LO has gained considerable interest. 12-LO activity results in the formation of 12-HPETE, which is reduced to 12-HETE. The reduction of 12-HPETE involves a glutathione-dependent peroxidase [118]. 12-LO has been partially characterized in mouse [119, 120], human [57, 121], rat [121] and guinea pig [122] epidermis and keratinocytes [92]. Furthermore, recently cDNA cloning of mouse epidermal 12-LOs has been carried out [120]. Both a cytosolic and a microsomal 12-LO were identified [120, 123]. Cytoplasmic 12-LO is also termed leukocyte-type lipoxygenase, whereas microsomal 12-LO is termed platelet-type lipoxygenase. Platelet-type 12-LO metabolizes AA exclusively into 12-HETE, whereas leukocyte-type 12-LO transforms AA into both 12-HETE and 15-HETE [120]. In normal human and mouse epidermis as well as in involved psoriatic epidermis and cultured human keratinocytes, only platelet-type 12-LO is detectable (mRNA and antibodies against the protein) [120, 124, 125]. However, in mouse epidermis both isozymes were induced transiently by phorbol esters [120]. Leukocyte-type 12-LO induced by phorbol ester treatment has been suggested to originate from Langerhans cells [120]. Both platelet-type and leukocyte-type 12-LO specifically result in the formation of 12(S)-HETE [120, 126]. It is therefore of interest that the 12-HETE in psoriatic scale is not the expected lipoxygenase-derived 12(S)-HETE but instead consists predominantly of 12(R)-HETE [127]. This stereochemical difference may indicate that 12-HETE is synthesized by different enzymes. For example, cytochrome P-450 monooxygenases have been reported to produce predominantly 12(R)-HETE rather than 12(S)-HETE from AA [128, 129]. Whether 12(R)-HETE recovered from psoriatic lesions is formed by the cytochrome P-450 system has not yet been clarified.

15-Lipoxygenase

The initial step of the 15-LO pathway is oxygenation of AA at C-15, resulting in formation of 15-HPETE, which can either be reduced by glutathione peroxidase to the corresponding 15-HETE or converted to 14, 15-LTA$_4$ by the action of LTA synthase. 14,15-LTA$_4$ is then transformed into 5, 15-diHETE, isomers of 8, 15-diHETE and isomers of 14,15-diHETE [130, 131]. Alternatively, 5,15-diHETE can be formed in a reaction catalyzed by 5-LO, because 15-HETE can act as substrate for 5-LO, leading to the formation of 5,15-diHETE [132]. 15-LO has been shown to be a Ca^{2+}-dependent enzyme localized predominantly in the cytosolic fraction [133]. 15-HETE formation may be increased by endogenous HETEs [134], and 15-HETE itself has been shown to be a potent activator of its own formation. Because 15-HETE is a potent 5-LO inhibitor [135], the increased formation of 15-HETE is most likely caused by inhibition of 5-LO, resulting in increased substrate availability for the 15-LO pathway.

15-LO has been shown to be present in a variety of tissues, including human skin [136] and human keratinocytes [137]. However, several careful studies have revealed rather conflicting data regarding 15-LO activity in human epidermis. 15-HETE was identified as the only mono-HETE formed by cultured human keratinocytes by several different groups [137–139], whereas others have found 12-HETE to be the prominent AA metabolite produced by homogeneous suspensions of freshly isolated epidermal cells, with only low amounts of 15-HETE present [92, 140]. Recently these differences have been explained by the various techniques used to isolate keratinocytes and epidermis. It has been demonstrated that AA metabolism by human keratinocytes depends on their functional differentiation. 12-HETE is mainly formed by the upper epidermal layers, whereas 15-HETE is formed predominantly by the basal keratinocytes [141].

Catabolism of leukotrienes

For cellular homeostasis and for limiting the proinflammatory effects of leukotrienes, it is essential that the leukotrienes generated in the skin are catabolized to less biologically active compounds. The metabolism of LTB_4 differs among tissues. In human epidermis LTB_4 is metabolized by an ω-oxidation pathway into 20-hydroxy-LTB_4 and 20-carboxy-LTB_4 [142, 143]. The first step of inactivation is the conversion of LTB_4 into 20-hydroxy-LTB_4 through the action of an NADPH-dependent microsomal cytochrome P-450 [142, 144]. The second step is oxidation of 20-hydroxy-LTB_4 into 20-carboxy-LTB_4 by an aldehyde dehydrogenase. Both 20-hydroxy-LTB_4 and 20-carboxy-LTB_4 are at least 10 times less potent than LTB_4 in inducing chemotaxis of guinea pig peritoneal eosinophils and neutrophils [145]. In cultured human keratinocytes LTB_4 has also been shown to be metabolized by a 12-hydroxydehydrogenase, resulting in the formation of 12-oxo-LTB_4, which is further metabolized into various dihydro-LTB_4 metabolites [146]. The LTB_4 12-hydroxydehydrogenase has been purified to homogeneity from porcine kidney cytosol [147], and very recently cDNA cloning of porcine and human LTB_4 12-hydroxydehydrogenase has been carried out [148]. 12-oxo-LTB_4 has been shown to be at least 100 times less potent than LTB_4 in increasing intracellular calcium [149].

Metabolic inactivation of peptideleukotrienes is thought to result from the transformation of LTC_4 to LTD_4 and LTE_4. However, both LTD_4 and LTE_4 have potent biological activities depending upon the biological system studied.

References

1   Willis AL (1981) Nutritional and pharmacological factors in eicosanoid biology. *Nutr Rev* 39: 289–301
2   Chapkin RS, Ziboh VA, Marcelo CL, Voorhees JJ (1985) Enzyme preparation from

human epidermis lack the capacity to transform linoleic acid (18:2, n6) and gamma-homolinoleic acid (20:3, n6) into arachidonic acid (20:4, n6). *J Lipid Res* 26: 348–353

3   Irvine RF (1982) How is the level of free arachidonic acid controlled in mammalian cells? *Biochem J* 204: 3–16

4   Ryborg AK, Deleuran B, Thestrup-Pedersen K, Kragballe K (1994) Lysophosphatidyl-choline: a chemotactic factor for human T-lymphocytes. *Arch Dermatol Res* 286: 6447–6451

5   Asaoka Y, Yoshida K, Sasaki Y, Nishizuka Y, Murakami M, Kudo I, Inoue K (1993) Possible role of mammalian secretory group II phospholipase A_2 in T-lymphocyte activation: Implication in propagation of inflammatory reaction. *Proc Natl Acad Sci USA* 90: 176–179

6   Dennis EA (1994) Diversity of group types, regulation and function of phospholipase A2. *J Biol Chem* 269: 13057–15060

7   Seilhammer JJ, Randall TL, Yamanaka M, Johnson LK (1986) Pancreatic phospholipase A_2: isolation of the gene and cDNAs from porcine pancrease and human lung. *DNA* 5: 519–527

8   Kramer RM, Hession C, Johansen B, Hayes P, McGray P, Chow EP, Tizard R, Pepinsky RB (1989) Structure and properties of a human non-pancreatic phospholipase A_2. *J Biol Chem* 264: 5768–5775

9   Clark JD, Milona N, Knopf JL (1990) Purification of a 110-kilo-dalton cytosolic phospholipase A_2 from the human monocytic cell line U937. *Proc Natl Acad Sci USA* 87: 7708–7712

10 Kudo I, Murakami M, Hara S, Inoue K (1993). Mammalian non-pancreatic phospholipase A_2. *Biochim Biophys Acta* 1170: 217–230

11 Andersen S, Sjursen W, Lægreid A, Volden G, Johansen B (1994) Elevated expression of human nonpancreatic phospholipase A_2 in psoriatic tissue. *Inflammation* 18: 1–12

12 McCord M, Chabot-Fletcher M, Breton, Marshall LA (1994) Human keratinocytes possess an *sn*-2 acylhydrolase that is biochemically similar to the U937-derived 85-kDa phospholipase A_2. *J Invest Dermatol* 102: 980–986

13 Davidson FF, Dennis EA (1990) Evolutionary relationships and implications for the regulation of phospholipase A_2 from snake venom to human secreted forms. *J Mol Evol* 31: 228–238

14 Hara S, Kudo I, Chang HW, Matsuta K, Miyamoto T, Inoue K (1989) Purification and characterization of extracellular phospholipase A_2 from human synovial fluid in rhumatoid arthritis. *J Biochem* 105: 395–399

15 Pruzanski W, Vadas P (1991) Phospholipase A_2 – a mediator between proximal and distal effectors of inflammation. *Immunol Today* 12: 143–146

16 Crowl RM, Stoller TJ, Conroy RR, Stoner CR (1991) Induction of phospholipase A_2 gene expression in human hepatoma cells by mediators of the acute phase response. *J Biol Chem* 266: 2647–2651

17 Muhl H, Geiger T, Pignat W, Marki F, van den Bosch H, Cerletti N, Cox D, McMaster G, Vosbeck K, Pfeilschifter J (1992) Transforming growth factors type-β and dexa-

methasone attenuate group II phospholipase A_2 gene expression by interleukin-1 and forskolin in rat mesangial cells. *FEBS Lett* 301: 190–194

18 Clark JD, Milona N, Knopf JL (1990) Purification of a 110-kilodalton cytosolic phospholipase A_2 from the human monocytic cell line U937. *Proc Natl Acad Sci USA* 87: 7708–7712

19 Leslie CC, Voelker DR, Channon JY, Wall MW, Zelarney PT (1988) Properties and purification of an arachidonyl-hydrolysing phospholipase A_2 from a macrophage cell line RAW 264.7. *Biochim Biophys Acta* 963: 476–492

20 Clark JD, Lin LL, Kriz WR, Ramesha CS, Sultzman LA, Lin AY, Milona N, Knopf JL (1991) A novel arachidonic acid-selective cytosolic PLA_2 contains a Ca^{2+}-dependent translocation domain with homology to PKC and GAP. *Cell* 65: 1043–1051

21 Channon JY, Leslie CC (1990) A calcium-dependent mechanism for associating a soluble arachidonyl-hydrolysing phospholipase A_2 with membrane in the macrophage cell line RAW 264.7. *J Biol Chem* 265: 5409–5413

22 Kim DK, Kudo I, Inoue K (1988) Detection in human platelets of phospholipase A_2 activity which preferentially hydrolyses an arachidonyl residue. *J Biochem* (Tokyo) 104: 492–494

23 Ramesha CS, Pickett WC (1986) Platelet activating factor and leukotriene biosynthesis is inhibited in polymorphonuclear leucocytes depleted of arachidonic acid. *J Biol Chem* 261: 7592–7595

24 Suga K, Kawasaki T, Blamk ML, Snyder F (1990) An arachidonoyl (polyenoic)-specific phospholipase A_2 activity regulates the synthesis of platelet activating factor in granulocytic HL-60 cells. *J Biol Chem* 265: 12363–12371

25 Murayama T, Ui M (1985) Receptor-mediated inhibition of adenylate cyclase and stimulation of arachidonic acid release in 3T3 fibroblasts. Selective susceptibility to islet-activating protein, pertussis toxin. *J Biol Chem* 260: 7226–7233

26 Burch RM, Luini A, Axelrod J (1986) Phospholipase A_2 and phospholipase C are activated by distinct GTP-binding proteins in response to β-adrenergic stimulation in FRTL5 thyroid cells. *Proc Natl Acad Sci USA* 83: 7201–7205

27 Pruzanski W, Vadas P, Fornasier V (1986) Inflammatory effect of intradermal administration of soluble phospholipase A_2 in rabbits. *J Invest Dermatol* 86: 380–382

28 Forster S, Ilderton E, Summerly R, Yardly HJ (1983) The level of phospholipase A_2 activity is raised in the uninvolved epidermis of psoriasis. *Br J Dermatol* 108: 103–105

29 Ryborg AK, Grøn B, Kragballe K (1995) Increased lysophosphatidylcholine content in lesional psoriatic skin. *Br J Dermatol* 134: 398–402

30 Ziboh VA, Lord JT (1979) Phospholipase A activity in the skin. Modulators of arachidonic acid release from phosphatidylcholine. *Biochem J* 184: 283–290

31 Kvedar J, Levine L (1987) Modulation of arachidonic acid metabolism in a cultured newborn keratinocyte cell line. *J Invest Dermatol* 88: 124–129

32 Hirata F (1981) The regulation of lipomodulin, a phospholipase inhibitory protein, in rabbit neutrophils by phosphorylation. *J Biol Chem* 256: 7730–7733

33 Ilchysyn A, Ilderton E, Kingsbury JA, Yardley HJ (1984) Evidence that raised levels of

phospholipase A_2 in uninvolved epidermis of psoriasis are caused by hyperphosphorylation of an inhibitor. *Br J Dermatol* 111: 721

34   Hong LS, Levine L (1976) Inhibition of arachidonic acid release from cells as the biochemical action of anti-inflammatory corticosteroids. *Proc Natl Acad Sci USA* 73: 1730–1734

35   Hirata F, Schiffermann E, Venkatasubramanian K, Salomon D, Axelrod J (1980) A phospholipase A_2 inhibitory protein in rabbit neutrophils induced by glucocorticoids. *Proc Natl Acad Sci USA* 77: 2533–2536

36   Norris JFB, Ilderton E, Yardley HJ, Summerly R, Forster S (1984) Utilization of epidermal phospholipase A_2 inhibition to monitor topical steroid action. *Br J Dermatol* 111 (suppl. 27): 195–203

37   Berridge MJ (1986) Cell signalling through phospholipid metabolism. *J Cell-Suppl* 4: 137–153

38   Hirata F, Axelrod J (1980) Phospholipid methylation and biological signal transmission. *Science* 209: 1082–1090

39   Rhee SG, Suh PG, Ryu SH, Lee SY (1989) Studies of inositol phospholipid-specific phospholipase. *Science* 244: 546–550

40   Bartel RL, Marcelo CL, Voorhees JJ (1987) Partial characterization of phospholipase C activity in normal, psoriatic uninvolved and lesional epidermis. *J Invest Dermatol* 88: 447–451

41   Cantley LL, Auger KR, Carpenter C, Duckworth B, Graziani A, Kapeller R, Soltoff S (1991) Oncogenes and signal transduction. *Cell* 64: 281–302

42   Cockroft S (1987) Phosphoinositide phosphodiesterase: regulation by a novel guanine nucleotide binding protein Gp. *Trends Pharmacol Sci* 12: 75–78

43   Nishibe S, Wahl MI, Hernandes-Sotomayor SMT, Tonks NK, Rhee SG, Carpenter G (1990) Increase of the catalytic activity of phospholipase C-γ_1 by tyrosine phosphorylation. *Science* 250: 1253–1256

44   Goldschmidt-Clermont PJ, Kim JW, Machesky LM, Rhee SG, Pollard TD (1991) Regulation of phospholipase C-g1 by profilin and tyrosine phosphorylation. *Science* 251: 1231–1233

45   Hamberg M, Svensson J, Wakabayashi T, Samuelsson B (1974) Isolation and structure of two prostaglandin endoperoxides that cause platelet aggregation. *Proc Natl Acad Sci USA* 71: 345–349

46   Ohki S, Ogino N, Yamamoto S, Hayaishi O (1979) Prostaglandin hydroperoxidase, an integral part of prostaglandin endoperoxide synthase from bovine vesicular gland microsomes. *J Biol Chem* 254: 829–836

47   Vane JR (1971) Inhibition of prostaglandin synthesis as a mechanism of action for aspirin-like drugs. *Nature New Biol* 231: 232–235

48   O'Banion MK, Sadowski HB, Winn V, Young DA (1991) A serum- and glucocorticoid-regulated 4-kilobase mRNA encodes a cyclooxygenase-related protein. *J Biol Chem* 266: 23261–23267

49   Kujubu DA, Fletcher BS, Varnum BC, Lim RW, Hirschmann HR (1991) TIS10, a phor-

bol ester tumor promotor-inducible mRNA from Swiss 3T3 cells, encodes a novel prostaglandin synthase/cyclooxygenase homologue. *J Biol Chem* 266: 12866–12872

50 Vane JR, Mitchell JA, Appleton I, Tomlinson A, Bishop-Bailey, Croxtall J, Willoughby DA (1994) Inducible isoforms of cyclooxygenase and nitric-oxide synthase in inflammation. *Proc Natl Acad Sci USA* 91: 2046–2050

51 Scholz K, Furstenberger G, Muller-Decker K, Marks F (1995) Differential expression of prostaglandin H synthase isozymes in normal and activated keratinocytes *in vivo* and *in vitro*. *Biochem J* 309: 263–269

52 Leong J, Hughes-Fulford M, Rakhlin N, Habib A, Maclouf J, Goldyne ME (1996) Cyclooxygenase in human and mouse skin and cultured human keratinocytes: association of COX-2 expression with human keratinocyte differentiation. *Exp Cell Res* 224: 79–87

53 Ogino N, Miyamoto T, Yamamoto S, Hayaishi O (1977) Prostaglandin endoperoxide E isomerase from bovine vesicular gland microsomes, a glutathione-requiring enzyme. *J Biol Chem* 252: 890–895

54 Jouvenaz GH, Nugteren DH, Beerthuis RK, van Dorp DA (1970) A sensitive method for the determination of prostaglandins by gas chromatography with electron-capture detection. *Biochim Biophys Acta* 202: 231–234

55 Ziboh VA, Hsia SL (1971) Prostaglandin E_2: biosynthesis and effects of glucose and lipid metabolism in rat skin. *Arch Biochem Biophys* 146: 100–109

56 Jonsson CE, Anggard E (1972) Biosynthesis and metabolism of prostaglandin E_2 in human skin. *Scand J Clin Lab Invest* 29: 289–296

57 Hammarström S, Lindgren JA, Marcello C, Duell EA, Anderson TE, Voorhees JJ (1979) Arachidonic acid transformations in normal and psoriatic skin. *J Invest Dermatol* 73: 180–183

58 Evans CB, Pillai SK, Goldyne ME (1993) Endogenous prostaglandin E_2 modulates calcium-induced differentiation in human skin keratinocytes. *Prostaglandins Leukot Essent Fatty Acids* 49: 777–781

59 Christ-Hazelhof E, Nugteren DH, van Dorp DA (1976) Conversions of prostaglandin endoperoxides by glutathione-S-transferases and serum albumins. *Biochim Biophys Acta* 450: 450–461

60 Christ-Hazelhof E, Nugteren DH (1979) Purification and characterization of prostaglandin endoperoxide D-isomerase, a cytoplasmic, gluthathione-requiring enzyme. *Biochim Biophys Acta* 572: 43–51

61 Ujihara M, Horiguchi Y, Ikai K, Urade Y (1988) Characterization and distribution of prostaglandin D synthetase in rat skin. *J Invest Dermatol* 90: 448–451

62 Ruzicka T, Printz MP (1982) Arachidonic acid metabolism in guinea pig skin. *Biochim Biophys Acta* 711: 391–397

63 Ruzicka T, Auböck J (1987) Arachidonic acid metabolism in guinea pig Langerhans cells: Studies on cyclooxygenase and lipoxygenase pathway. *J Immunol* 138: 539–543

64 Watanabe K, Iguchi Y, Iguchi S, Arai Y, Hayaishi O, Roberts LJ (1987) Stereospecific conversion of prostaglandin D_2 to $9\alpha,11\beta$-prostaglandin F_2 and prostaglandin H_2 to

prostaglandin $F_{2\alpha}$ by PGF synthase. *Adv Prostaglandin Thromboxane Leukot Res* 17A: 44–46

65 Chang DGB, Sun M, Tai HH (1981) Prostaglandin 9-ketoreductase and type II 15-hydroxyprostaglandin dehydrogenase from swine kidney are alternate activities of a single enzyme protein. *Biochem Biophys Res Commun* 99: 745–751

66 Reingold DF, Kawasaki A, Needleman P (1981) A novel prostaglandin 11-keto reductase found in rabbit liver. *Biochim Biophys Acta* 659: 179–188

67 Ziboh VA, Lord JT, Penneys NS (1977) Alterations of prostaglandin E_2-9-ketoreductase activity in proliferating skin. *J Lipid Res* 18: 37–43

68 Anderson MW, Crutchley DJ, Tainer BE, Eling TE (1978) Kinetic studies on the conversion of prostaglandin endoperoxide PGH_2 by thromboxane synthase. *Prostaglandins* 16: 563–570

69 Salmon JA, Smith DR, Flower RJ, Moncada S, Vane JR (1978) Further studies on the enzymatic conversion of prostaglandin endoperoxide into prostacyclin by procine aorta microsomes. *Biochim Biophys Acta* 523: 250–262

70 Rouzer CA, Matsumoto T, Samuelsson B (1986) Single protein from human leukocytes possesses 5-lipoxygenase and LTA_4 synthase activities. *Proc Natl Acad Sci USA* 83: 857–861

71 Rouzer CA, Kargman S (1988) Translocation of 5-lipoxygenase to the membrane in human leukocytes challenged with ionophore A23187. *J Biol Chem* 263: 10980–10988

72 Wong A, Hwang SM, Cook MN, Hogaboom GK, Crooke ST (1988) Interactions of 5-lipoxygenase with membranes: studies on the associations of soluble enzyme with membranes and alterations in enzyme activity. *Biochemistry* 27: 6763–6769

73 Miller DK, Gillard JW, Vickers PJ, Sadowski S, Léveillé C, Mancini JA, Charleson P, Dixon RAF, Ford-Hutchinson AW, Fortin R et al (1990) Identification and isolation of a membrane protein necessary for leukotriene production. *Nature* 343: 278–281

74 Dixon RAF, Diehl RE, Opas E, Rands E, Vickers PJ, Evans JF, Gillard JW, Miller DK (1990) Requirement of a 5-lipoxygenase-activating protein for leukotriene synthesis. *Nature* 343: 282–284

75 Shimizu T, Rådmark O, Samuelsson B (1984) Enzyme with dual lipoxygenase activities catalyzes leukotriene A_4 synthesis from arachidonic acid. *Proc Natl Acad Sci USA* 81: 689–693

76 Borgeat P, Hamberg M, Samuelsson B (1976) Transformation of arachidonic acid and homo-gamma-linolenic acid by rabbit polymorphonuclear leukocytes: monohydroxy acids from novel lipoxygenases. *J Biol Chem* 251: 7816–7820

77 Lewis RA, Austen KF, Soberman RJ (1990) Leukotrienes and other products of the 5-lipoxygenase pathway. Biochemistry and relation to pathobiology in human diseases. *N Engl J Med* 323: 645–655

78 Rådmark O, Malmsten C, Samuelsson B (1980) Leukotriene A_4: enzymatic conversion to leukotriene C_4. *Biochem Biophys Res Commun* 96: 1679–1687

79 Rådmark O, Malmsten C, Samuelsson B, Goto G, Marfat A, Corey EJ (1980)

Leukotriene A. Isolation from human polymorphonuclear leukocytes. *J Biol Chem* 255: 11828–11831

80 Maycock AL, Anderson MS, DeSousa DM, Kuehl FA Jr (1982) Leukotriene A_4: Preparation and enzymatic conversion in a cell-free system to leukotriene B_4. *J Biol Chem* 257: 13911–13914

81 Rådmark O, Shimizu T, Jörnvall H, Samuelsson B (1984) Leukotriene A_4 hydrolase in human leukocytes. Purification and properties. *J Biol Chem* 259: 12339–12345

82 Jakschik BA, Kuo CG (1983) Characterization of leukotriene A_4 and B_4 biosynthesis. *Prostaglandins* 25: 767–782

83 Sun FF, McGuire JC (1984) Metabolism of arachidonic acid by human neutrophils. Characterization of the enzymatic reactions that lead to the synthesis of leukotriene B_4. *Biochim Biophys Acta* 794: 56–64

84 Orning L, Hammarström S (1980) Inhibition of leukotriene C and leukotriene D biosynthesis. *J Biol Chem* 255: 8023–8026

85 Lee CW, Lewis RA, Corey EJ, Austen KF (1983) Conversion of leukotriene D_4 to leukotriene E_4 by dipeptidase released from specific granule of human polymorphonuclear leukocytes. *Immunology* 48: 27–35

86 Brain S, Camp R, Dowd P, Black AK, Greaves M (1984) The release of leukotriene B_4-like material in biologically active amounts from lesional skin of psoriasis. *J Invest Dermatol* 83: 70–73

87 Grabbe J, Czarnetzki MB, Rosenbach T, Mardin M (1984) Identification of chemotactic lipoxygenase products of arachidonate metabolism in psoriasis. *J Invest Dermatol* 82: 477–479

88 Brain SD, Camp RDR, Kobza Black A, Dowd PM, Greaves MW, Ford-Hutchinson AW, Charleson S (1985) Leukotriene C_4 and D_4 in psoriatic skin lesions. *Prostaglandins* 29: 611–619

89 Sampson AP, Thomas RU, Costello JF (1992) Enhanced leukotriene synthesis in leukocytes of atopic dermatis and asthma subjects. *Br J Clin Pharmacol* 33: 423–430

90 Grabbe J, Rosenbach T, Czarnetzki BM (1985) Production of LTB_4-like chemotactic arachidonate metabolites from human keratinocytes. *J Invest Dermatol* 85: 527–530

91 Rosenbach T, Grabbe J, Moller A, Schwanitz HJ, Czarnetski BM (1985) Generation of leukotrienes from normal epidermis and their demonstration in cutaneous disease. *Br J Dermatol* 113 (suppl 28): 157–167

92 Ziboh VA, Casebolt TL, Marcelo CL, Voorhees JJ (1984) Lipoxygenation of arachidonic acid by subcellular preparations from murine keratinocytes. *J Invest Dermatol* 83: 248–251

93 Janssen-Timmen U, Vickers PJ, Wittig U, Lehmann WD, Stark HJ, Fusenig NE, Rosenbach T, Rådmark O, Samuelsson B, Habenicht AJR (1995) Expression of 5-lipoxygenase in differentiating human skin keratinocytes. *Proc Natl Acad Sci USA* 92: 6966–6970

94 Brenton J, Woof D, Young P, Chabot-Fletcher M (1996) Human keratinocytes lack the components to produce leukotriene B_4. *J Invest Dermatol* 106: 162–167

95   Iversen L, Fogh K, Ziboh VA, Kristensen P, Schmedes A, Kragballe K (1993) Leukotriene B_4 formation during human neutrophil keratinocyte interactions: evidence for transformation of leukotriene A_4 by putative keratinocyte leukotriene A_4 hydrolase. *J Invest Dermatol* 100: 293–298

96   Iversen L, Kristensen P, Grøn B, Ziboh VA, Kragballe K (1994) Human epidermis transforms exogenous leukotriene A_4 into peptide leukotrienes: possible role in transcellular metabolism. *Arch Dermatol Res* 286: 261–267

97   Iversen L, Ziboh VA, Shimizu T, Ohishi N, Rådmark O, Wetterholm A, Kragballe K (1994) Identification and subcellular localization of leukotriene A_4-hydrolase activity in human epidermis. *J Dermatol Sci* 7: 191–201

98   Solá J, Godessart N, Vila L, Puig L, de Moragas JM (1992) Epidermal cell-polymorphonuclear leukocyte cooperation in the formation of leukotriene B_4 by transcellular biosynthesis. *J Invest Dermatol* 98: 333–339

99   Iversen L, Deleuran B, Hoberg AM, Kragballe K (1996) LTA_4 hydrolase in human skin: decreased activity, but normal concentration in lesional psoriatic skin. Evidence for different LTA_4 hydrolase activity in human lymphocytes and human skin. *Arch Dermatol Res* 288: 217–224

100   Iversen L, Kristensen P, Nissen JB, Merrick WC, Kragballe K (1995) Purification and characterization of leukotriene A_4 hydrolase from human epidermis. *FEBS Lett* 358: 316–322

101   Nissen JB, Iversen L, Kragballe K (1995) Characterization of the aminopeptidase activity of epidermal leukotriene A_4 hydrolase against the opioid dynorphin fragment 1–7. *Br J Dermatol* 133: 742–749

102   Griffin KJ, Gierse J, Krivi G, Fitzpatrick FA (1992) Opioid peptides are substrates for the bifunctional enzyme LTA_4 hydrolase/aminopeptidase. *Prostaglandins* 44: 251–257

103   Haeggström J, Bergman T, Jörnvall H, Rådmark O (1988) Guinea-pig liver leukotriene A_4 hydrolase. Purification, characterization and structural properties. *Eur J Biochem* 174: 717–724

104   Orning L, Jones DA, Fitzpatrick FA (1990) Mechanism-based inactivation of leukotriene A_4 hydrolase during leukotriene B_4 formation by human erythrocytes. *J Biol Chem* 265: 14911–14916

105   Sala A, Bolla M, Zarini S, Müller-Peddinghaus R, Folco G (1996) Release of leukotriene A_4 versus leukotriene B_4 from human polymorphonuclear leukocytes. *J Biol Chem* 271: 17944–17948

106   Söderström M, Hammarström S, Mannervik B (1988) Leukotriene C synthase in mouse mastocytoma cells. An enzyme distinct from cytosolic and microsomal glutathione transferases. *Biochem J* 250: 713–718

107   Raza H, Awasthi YC, Zaim MT, Eckert RL, Mukhtar H (1991) Glutathione S-transferases in human and rodent skin: multiple forms and species-specific expression. *J Invest Dermatol* 96: 463–467

108   Agarwal R, Raza H, Allyn DL, Bickers DR, Mukhtar H (1992) Glutathione S-trans-

ferase-dependent conjugation of leukotriene A_4-methyl ester to leukotriene C_4-methyl ester in mammalian skin. *Biochem Pharmacol* 44: 2047–2053

109 Meijer L, Brash AR, Bryant RW, Machlouf J, Sprecher H (1986) Stereospecific induction of starfish oocyte maturation by (8R)-hydroxyeicosatetraenoic acid. *J Biol Chem* 261: 17040–17047

110 Goetzl EJ, Sun FF (1979) Generation of unique mono-hydroxy-eicosatetraenoic acids from arachidonic acid by human neutrophils. *J Exp Med* 150: 406–411

111 Hunter JA, Finkbeiner WE, Nadel JA, Goetzl EJ (1985) Predominant generation of 15-lipoxygenase metabolites of arachidonic acid by epithelial cells from human trachea. *Proc Natl Acad Sci USA* 82: 4633–4637

112 Camp RDR, Mallet AJ, Woollard PM, Brain SD, Black AK, Greaves MW (1983) The identification of hydroxy fatty acids in psoriatic skin. *Prostaglandins* 26: 431–447

113 Hughes MA, Brash AR (1991) Investigation of the mechanism of biosynthesis of 8-hydroxyeicosatetraenoic acid in mouse skin. *Biochim Biophys Acta* 1081: 347–354

114 Fürstenberger G, Hagedorn H, Jacobi T, Besemfelder E, Stephan M, Lehmann WD, Marks F (1991) Characterization of an 8-lipoxygenase activity induced by the phorbol ester tumor promoter 12-O-tetradecanoylphorbol-13-acetate in mouse skin *in vivo*. *J Biol Chem* 266: 15738–15745

115 Forman BM, Chen J, Evans RM (1997) Hypolipidemic drugs, polyunsaturated fatty acids and eicosanoids are ligands for peroxisome proliferator-activated receptors α and δ. *Proc Natl Acad Sci USA* 94: 4312–4317

116 Kliewer SA, Sundseth SS, Jones SA, Brown PJ, Wisely GB, Koble CS, Devchand P, Wahli W, Willson TM, Lenhard JM et al (1997) Fatty acids and eicosanoids regulate gene expression through direct interactions with peroxisome proliferator-activated receptors α and γ. *Proc Natl Acad Sci USA* 94: 4318–4323

117 Hammarström S, Hamberg M, Samuelsson B, Duell EA, Stawiski M, Voorhees JJ (1975) Increased concentrations of nonesterified arachidonic acid, 12L-hydroxy-5,8,10,14-eicosatetraenoic acid, prostaglandin E_2, prostaglandin $F_{2\alpha}$ in epidermis of psoriasis. *Proc Natl Acad Sci USA* 72: 5130–5134

118 Chang WC, Nakao J, Orimo H, Murota SI (1982) Effects of reduced glutathione on the 12-lipoxygenase pathways in platelets. *Biochem J* 202: 771–776

119 Henke D, Danilowicz R, Eling T (1986) Arachidonic acid metabolism by isolated epidermal basal and differentiated keratinocytes from the hairless mouse. *Biochim Biophys Acta* 876: 271–279

120 Krieg P, Kinzig A, Ress-Löschke M, Vogel S, Vanlandigham B, Stephan M, Lehmann WD, Marks F, Fürstenberger G (1995) 12-Lipoxygenase isozymes in mouse skin tumor development. *Mol Carcinog* 14: 118–129

121 Nugteren DH, Kivits GAA (1987) Conversion of linoleic acid and arachidonic acid by skin epidermal lipoxygenases. *Biochim Biophys Acta* 921:135–141

122 Ruzicka T, Vitto A, Printz MP (1983) Epidermal arachidonate lipoxygenase. *Biochim Biophys Acta* 751: 369–374

123 Nakadate T, Aizu E, Yamamoto S, Kato R (1986) Some properties of lipoxygenase activ-

ities in cytosol and microsomal fractions of mouse epidermal homogenate. *Prostaglandins Leukot Med* 21: 305–309

124 Hussain H, Shornick LP, Shannon VR, Wilson JD, Funk CD, Pentland AP, Holtzman MJ (1994) Epidermis contains platelet-type 12-lipoxygenase that is overexpressed in germinal layer keratinocytes in psoriasis. *Am J Physiol* 266: C243–253

125 Takahashi Y, Reddy GR, Ueda N, Yamamoto S, Arase S (1993) Arachidonate 12-lipoxygenase of platelet-type in human epidermal cells. *J Biol Chem* 268:16443–16448

126 Hamberg M, Samuelsson B (1974) Prostaglandin endoperoxides. Novel transformations of arachidonic acid in human platelets. *Proc Natl Acad Sci USA* 71: 3400–3404

127 Wollard PM (1986) Stereochemical difference between 12-hydroxy-5,8,10,14-eicosatetraenoic acid in platelets and psoriatic lesions. *Biochem Biophys Res Commun* 136: 169–176

128 Schwartzman ML, Balazy M, Masferrer J, Abraham NG, McGiff JC, Murphy RC (1987) 12(R)-Hydroxyeicosatetraenoic acid: a cytochrome P450-dependent arachidonate metabolite that inhibits Na, K-ATPase in the cornea. *Proc Natl Acad Sci USA* 84: 8125–8129

129 Capdevila J, Yadagiri P, Manna S, Falck JR (1986) Absolute configuration of the hydroxyeicosatetraenoic acids (HETEs) formed during catalytic oxygenation of arachidonic acid by microsomal cytochrome P-450. *Biochem Biophys Res Commun* 141: 1007–1011

130 Maas RL, Turk J, Oates JA, Brash A (1982) Formation of a novel dihydroxy acid from arachidonic acid by lipoxygenase-catalyzed double oxygenation in rat mononuclear cells and human leukocytes. *J Biol Chem* 257: 7056–7067

131 Turk J, Maas RL, Brash AR, Roberts LJ, Oates JA (1982) Arachidonic acid 15-lipoxygenase products from human eosinophils. *J Biol Chem* 257: 7068–7076

132 Wilkinson D, Hallam C, Hamsley PE, Lord GH, Mitchell PD (1985) 15-L(S)-hydroxy-5z, 8z, 11z, 13e-eicosatetraenoic acid is a substrate for 5-lipoxygenase. *Biochem Soc Trans* 13: 180–182

133 Burrall BA, Wintroub BU, Goetzl EJ (1986) Cytosolic localization of the 15-lipoxygenase of human neonatal foreskin keratinocytes. *J Invest Dermatol* 86: 466 (abstr)

134 Vanderhoek JY, Karmin HT, Ekborg SL (1985) Endogenous hydroxyeicosatetraenoic acids stimulate the human polymorphonuclear leukocyte 15-lipoxygenase pathway. *J Biol Chem* 260: 14482–14487

135 Vanderhoek JY, Bryant RW, Baily JM (1980) Inhibition of leukotriene biosynthesis by the leukocyte product 15-hydroxy-5, 8, 11, 13-eicosatetraenoic acid. *J Biol Chem* 255: 10064–10066

136 Kragballe K, Pinnamaneni G, Desjarlais L, Duell EA, Voorhess JJ (1986) Dermis-derived 15-hydroxy-eicosatetraenoic acid inhibits epidermal 12-lipoxygenase activity. *J Invest Dermatol* 87: 494–498

137 Burrall BA, Wintroub BU, Goetzl EJ (1985) Selective expression of 15-lipoxygenase activity by cultured human keratinocytes. *Biochem Biophys Res Commun* 133: 208–213

138 Burall BA, Cheung M, Chiu A, Goetzl EJ (1988) Enzymatic properties of the 15-lipoxy-genase of human cultured keratinocytes. *J Invest Dermatol* 71: 157–162

139 Green FA (1989) Generation of lipoxygenase products in normal and membrane-damaged cultured human keratinocytes. *J Invest Dermatol* 93: 486–491

140 Holtzman MJ, Turk J, Pentland A (1989) A regiospecific monooxygenase with novel stereopreference is the major pathway for arachidonic acid oxygenation in isolated epidermal cells. *J Clin Invest* 84: 1446–1453

141 von Zepelin HHH, Schröder JM, Smid P, Reusch MK, Christophers E (1991) Metabolism of arachidonic acid by human epidermal cells depends upon maturational stage. *J Invest Dermatol* 97: 291–291

142 Mukhtar H, Bik DP, Ruzicka T, Merk HF, Bickers DR (1989) Cytochrome P-450-dependent omega-oxidation of leukotriene B_4 in rodent and human epidermis. *J Invest Dermatol* 93: 231–235

143 Morelli JG, Norris DA, Lyons MB, Murphy RC (1988) Metabolism of exogenous leukotrienes by cultured human keratinocytes and melanocytes. *J Invest Dermatol* 90: 590

144 Sumimoto J, Takeshige K, Minakami S (1988) Characterization of human neutrophil leukotriene B_4 omega-hydroxylase as a system involving a unique cytochrome P-450 and NADPH-cytochrome P-450 reductase. *Eur J Biochem* 172: 315–324

145 Ford-Hutchinson AW, Rackham A, Zamboni R, Rokach J, Roy S (1983) Comparative biological activities of synthetic leukotriene B_4 and its omega-oxidation products. *Prostaglandins* 25: 29–37

146 Wheelan P, Zirolli JA, Morelli JG, Murphy RC (1993) Metabolism of leukotriene B_4 by cultured keratinocytes. *J Biol Chem* 268: 25439–25448

147 Yokomizo T, Izumi T, Takahashi T, Kasama T, Kobayashi Y, Sato F, Taketani Y, Shimizu T (1993) Enzymatic inactivation of leukotriene B_4 by a novel enzyme found in the porcine kidney. Purification and properties of leukotriene B_4 12-hydroxydehydrogenase. *J Biol Chem* 268: 18128–18135

148 Yokomizo T, Ogawa Y, Uozumi N, Kume K, Izumi T, Shimizu T (1996) cDNA cloning, expression and mutagenesis study of leukotriene B_4 12-hydroxydehydrogenase. *J Biol Chem* 271: 2844–2850

149 Yokomizo T, Uozumi N, Takahashi T, Kume K, Izumi T, Shimizu T (1995) Leukotriene A_4 and leukotriene B_4 metabolism. *J Lipid Cell Signalling* 12: 321–332

Role of eicosanoids in psoriasis and atopic skin diseases

Karsten Fogh and Knud Kragballe

Department of Dermatology, Marselisborg Hospital, University of Aarhus,
DK-8000 Aarhus C, Denmark

Introduction

As described in detail in the chapter on arachidonic acid metabolism in the skin, eicosanoids are formed after oxygenation of arachidonic acid. Prostaglandins are formed by the cyclooxygenase (CO) pathway, whereas leukotrienes (LTs) and hydroxyeicosatetraenoic acids (HETEs) are formed by the 5-, 12- and 15-lipoxygenase (LO) pathways. 5- and 12-LO products are mainly proinflammatory in the skin, whereas the main 15-LO product 15-HETE has antiinflammatory capacities.

During the past decade eicosanoids have been assigned a role in inflammatory skin diseases such as psoriasis and atopic dermatitis because of their biological action in the skin and their presence in bioactive amounts in inflamed skin lesions. This chapter reviews the potential role of these compounds in the pathophysiology of psoriasis and atopic dermatitis.

Biological effects of eicosanoids in human skin

In vitro LTB_4 has several proinflammatory characteristics. It is a very potent chemotactic and chemokinetic agent with effects in the concentration range of 10^{-8}–10^{-11} M [1]. In neutrophils LTB_4 causes degranulation, [2] cation fluxes [3] and enhanced binding to endothelial cells, [4] and most of these events are mediated via specific LTB_4 receptors [5]. In addition to the proinflammatory effects, mediated through the effect on leukocytes, LTB_4 also has the capacity to stimulate DNA synthesis in cultured human keratinocytes [6]. LTB_4 has also been demonstrated to induce pigmentation in human melanocytes [7, 8], which possibly plays a role in postinflammatory pigmentation. The biological effects of LTB_4 are highly stereospecific to LTB_4 [9] because isomers of LTB_4 [10] and ω-oxidation products [11, 12] possess less biological activity than LTB_4. Although 20-OH-LTB_4 still retains significant biological activity, 20-COOH-LTB_4 is virtually inactive *in vivo* and has less than 1% of the activity of LTB_4 *in vitro* [11]. These data suggest that although ω-oxidation of

LTB$_4$ represents an inactivation pathway, the metabolites formed have some biological activity. *In vitro* activities of eicosanoids are summarized in Table 1.

In vivo LTB$_4$ has been demonstrated to induce skin inflammation. Topical application under occlusion of microgram amounts of LTB$_4$ to normal human skin results in a dose-dependent wheal and flare reaction lasting several days [13]. Histologically, this reaction is characterized by epidermal neutrophil microabscesses, which have been shown to be maximal after 24 h and to be accompanied by neutrophil and mononuclear dermal infiltrates [13, 14]. Also, intraepidermal injection of LTB$_4$ elicits a transient wheal and flare reaction as well as accumulation of neutrophils perivascularly [15]. Concomitant injection of LTB$_4$ and prostaglandin E$_2$ (PGE$_2$) into human skin results in amplification of the inflammatory response [16]. The accumulation of neutrophils is an important component of the inflammatory response to LTB$_4$. Therefore, the capacity of neutrophils to respond to LTB$_4$ will largely determine the importance of LTB$_4$ as an inflammatory mediator. *In vitro* prior exposure of neutrophils to LTB$_4$ results in a specific loss of responsiveness to LTB$_4$ [17]. This desensitization may explain why repeated topical applications of LTB$_4$ to normal human skin result in a diminished inflammatory response [18].

Topical application of LTB$_4$ also leads to epidermal hyperproliferation [19], probably mediated by a direct mitogenic effect which has been demonstrated in cul-

Table 1 - Biologic effect of lipoxygenase products in vitro

Effect	Metabolites
Leukocyte migration (chemotaxis, chemokinesis)	LTB$_4$, 5-HETE, 12(R)-HETE, 12(S)-HETE
Aggregation Degranulation Superoxide generation Leukocyte/endothelium adhesion	LTB$_4$
5-LO inhibition	15-HETE, 13-HODD
12-LO inhibition	15-HETE
Inhibition of LTB$_4$-induced neutrophil chemotaxis	15-HETE
Smooth muscle contraction	LTC$_4$, LTD$_4$
Increase in DNA synthesis (human keratinocytes)	LTB$_4$, 12-HETE, LTC$_4$, LTD$_4$
Stimulation of T-suppressor cells Inhibition of T-helper cells	LTB$_4$, 15-HETE
Enhancement of basophil histamine release	LTB$_4$, 5-HPETE, 5-HETE, 12-HPETE

tured human keratinocytes [6]. This is also supported by the identification a specific LTB$_4$-receptor on cultured human keratinocytes [20].

Also the peptido-leukotrienes (LTC$_4$, LTD$_4$ and LTE$_4$) increase DNA synthesis in cultured human keratinocytes [6]. The peptido-LTs are potent vasodilators and have potent actions on smooth muscle. Species- and tissue-related differences in the effects of the peptido-leukotrienes have been described. However, *in vivo* data obtained in human skin have shown that peptido-leukotrienes cause increased vascular permeability [21], vasodilation [22] and produce a wheal and flare reaction [23]. Like LTB$_4$, the peptido-leukotrienes are extremely potent, with LTC$_4$ as the most potent and LTE$_4$ as the least potent. Based upon their *in vitro* and *in vivo* activities these potent mediators are thought to play an important role in allergic and immediate hypersensitivity reactions. In further support of this assumption is the release of peptido-LTs after challenge with specific allergen [24, 25].

Another proinflammatory eicosanoid is the 12-LO product 12-HETE. As mentioned in the chapter on eicisanoid metabolism, 12(R)-HETE is predominantly produced in the skin and is also found in psoriatic lesions [26]. 12(R)-HETE is more potent in inducing skin inflammation than 12(S)-HETE [27]. *In vitro* 12-HETE has some effects in common with LTB$_4$. It stimulates DNA synthesis in cultured human keratinocytes [28] and it induces neutrophil chemotaxis [29], although 12-HETE is a much less potent chemoattractant than LTB$_4$. *In vivo* the cutaneous response to topical application includes induction of erythema and accumulation of neutrophils and mononuclear leukocytes [30].

The CO product of AA, PGE$_2$, has also been shown to have proinflammatory properties. Intradermal injections of PGE$_2$ (1–10 mg) into human skin causes erythema [31], with higher doses also producing edema [32]. Furthermore, PGE$_2$ can amplify the vascular permeability changes induced by other mediators [32, 33]. These responses appear to be due to a direct effect of PGE$_2$ on blood vessels. Specific and high-affinity binding sites for PGE$_2$ have been located on the membrane fraction of human epidermal cells [34]. The proinflammatory properties of PGE$_2$ may be balanced by its potential immunosuppressive effects. PGE$_2$ inhibits T cell proliferation, lymphocyte chemotaxis and interleukin-1 (IL-1) formation [35–37].

In contrast to PGE$_2$, 12-HETE and the leukotrienes, the 15-LO product of AA, 15-HETE, has antiinflammatory properties. *In vitro* 15-HETE has been shown to inhibit LTB$_4$ formation [38] and 12-HETE formation [39], and it specifically inhibits the neutrophil chemotactic effect of LTB$_4$ [40]. The inhibition of LTB$_4$ formation is probably due to modulation of 5-LO, because no changes in PGE$_2$ formation have been determined [41]. *In vivo*, 15-HETE inhibits LTB$_4$-induced erythema and edema [42], and reduces LTB$_4$ in the synovial fluid of carragheenan-induced experimental arthritis in dogs [43]. 15-HETE also has some immunomodulatory effects. It inhibits the mixed lymphocyte reaction [44], induces generation of murine cytotoxic suppressor T cells [45], and decreases γ-interferon production by murine lymphoma cells [46]. Furthermore, IL-4 and IL-13 have

Table 2 - Biological effect of lipoxygenase products in vivo

Effect	Metabolites
Wheal and flare	LTC_4, LTD_4, LTB_4
Vasodilatation	LTC_4, LTD_4
Increase in vascular permeability	LTC_4, LTD_4, LTB_4
Chemotaxis	LTB_4
Intraepidermal microabscesses Epidermal hyperproliferation	LTB_4, 12-HETE

recently been shown to be potent activators of 15-LO in mononuclear cells [47–49]. *In vivo* activities of eicosanoids are summarized in Table 2.

Eicosanoids in psoriasis

Psoriasis is a chronic hyperinflammatory and hyperproliferative disease occurring in about 2% of the Western population. The clinical appearance of the disease is characterized by sharply defined symmetrically located plaques of varying size from pinpoint to large plaques. The lesions are infiltrated, scaly and erythematous. Guttate psoriasis is characterized by small lesions, whereas the term 'chronic plaque psoriasis' is used for coin-sized to palm-sized plaques.

Histologically, psoriatic lesions are characterized by hyperplasia and impaired differentiation of the epidermal keratinocytes, by dilated capillaries in the papillary dermis, and by leukocytes infiltrating both epidermis and dermis. The psoriatic epidermis is characterized by increased multiplication and turnover of keratinocytes and decreased keratinocyte differentiation. Furthermore, microabscesses resulting from the accumulation of neutrophils having migrated into the epidermis are typical findings. The dermal changes consist of excess endothelial gaps, hyperproliferation of the endothelial cells of the postcapillary venules.

As mentioned above, psoriasis is characterized by tightly associated inflammatory and proliferative processes. As yet it is unknown which factors are responsible for these changes. Because certain LO products (LTB_4, LTC_4, LTD_4 and 12-HETE) have both chemoattractant and mitogenic potency in the skin, they may be potential mediators of inflammation and disturbed cell growth in psoriasis. In contrast to the proinflammatory eicosanoids, 15-HETE has the capacity to inhibit formation of LTB_4 and 12-HETE and the chemotactic effect of LTB_4. Thus 15-HETE may be an important regulator of inflammation induced by LTB_4 and/or 12-HETE.

Both CO products and LO products have proinflammatory properties. However, with respect to psoriasis the LO products are of particular interest, and the leukotrienes especially have been investigated for their pathophysiological role in the disease. In support of the idea of leukotrienes playing a central role in the pathogenesis of psoriasis are the following: LTB_4 [50–52] and peptido-leukotrienes [53] have been extracted from psoriatic skin lesions, and LTB_4 at least appears to be present in biologically active amounts [50]. Furthermore, a significant increase in urinary LTE_4 has been found in psoriatic patients compared with healthy human volunteers [54]. Single topical applications of LTB_4 to normal human skin result in the formation of intraepidermal microabscesses [13, 14], which is one of the earliest and characteristic morphologic events in psoriasis. LTB_4 stimulates human epidermal keratinocyte proliferation both *in vitro* and *in vivo* [19]. In most studies where LTB_4 was determined in psoriatic skin, chronic plaques were analyzed [50–52]. However, biologically active amounts of LTB_4, but not other eicosanoids, are also detectable in acute, guttate lesions, supporting a role for LTB_4 in the early stages of psoriasis [55]. Even though these results strongly indicate that leukotrienes are involved in the pathophysiology of psoriasis, this has recently been questioned [56], partly because the actual cellular origin of LTB_4 has not been definitely established yet. Low quantities of LTB_4 formation have been reported in freshly isolated human epidermal cells [57, 58] and cultured mouse keratinocytes [59], as determined by high-pressure liquid chromatography (HPLC) [59] and by radioimmunoassay (RIA) and chemotactic activity [57, 58]. However, we and others have not been able to determine any 5-LO activity in cultured human keratinocytes and human epidermis, and it is, therefore, possible that leukotriene formation in the skin may not be the initial event that causes neutrophils to migrate into the skin in the developing psoriatic plaque. Neutrophil migration may be initiated by other chemoattractants such as platelet-activating factor, C5a and IL-8, and leukotriene formation may be a secondary event amplifying the progressing inflammatory process. This hypothesis is in accordance with the idea of transcellular leukotriene synthesis as a major pathway for leukotriene synthesis in inflammatory skin diseases.

We and others have recently reported that coincubation of human neutrophils and cultured human keratinocytes as well as human neutrophils and human epidermis results in increased formation of LTB_4 and LTC_4 compared with incubation of these cell types separately [60–63]. Furthermore, it was shown that the increase in leukotriene formation was a result of transformation of neutrophil-derived LTA_4 in the keratinocytes.

In psoriasis the neutrophils are in close contact with the keratinocytes because of the migration into the epidermis. Therefore, the epidermis may play an active role in the formation of proinflammatory mediators in this disease. A key enzyme in transcellular LTB_4 synthesis is epidermal LTA_4 hydrolase. This enzyme has been demonstrated in the epidermis by activity determination [60, 61, 63–65], by immunohistochemical staining of both normal and psoriatic skin [64, 65], and by purification from normal human epidermis and cultured human keratinocytes [66]. A characteristic feature of LTA_4 hydrolase is that it undergoes suicide inactivation while transforming LTA_4 into LTB_4 [66, 67]. This inactivation is due to covalent binding of substrate LTA_4 to LTA_4 hydrolase. Suicide inactivation of the LTA_4 hydrolase may be part of an autoregulation of LTB_4 formation. However, when neutrophils migrate into the epidermis, a marked increase in LTA_4 hydrolase is accessible, and thereby the self-limitation in neutrophil LTB_4 synthesis is bypassed. It is of related interest that LTA_4 hydrolase activity expressed as nanograms of LTB_4 formed per milligram of LTA_4 hydrolase when incubated with exogenous LTA_4 is significantly decreased in involved psoriatic skin compared with matched uninvolved psoriatic skin [64], indicating that transcellular LTB_4 synthesis has taken place in the involved psoriatic skin.

From the above data it can be concluded that LTB_4 in particular may play an important role in the pathogenesis of psoriasis because of its biological profile and its presence in the lesional skin at biologically active concentrations. Formation of LTB_4 in lesional skin is most likely a result of transcellular metabolism, where LTA_4 released from activated leukocytes infiltrating the epidermis is transformed into LTB_4 by LTA_4 hydrolase. Thus, LTA_4 hydrolase seems to be a key enzyme in the formation of LTB_4 in the skin as recently reviewed by us [68], suggesting that inhibitors of LTA_4 hydrolase may be used in the treatment of psoriasis.

Cyclosporin A, an immunosuppressive drug known to have an antipsoriatic effect, has previously been shown to decrease LTB_4 content in lesional psoriasis [69]. The mechanism of this LTB_4 decrease is not known. Recently, we reported that incubation of cultured human keratinocytes with cyclosporin A for 72 h results in downregulation of the LTA_4 hydrolase content in these cells [70]. This may be one of the mechanisms by which cyclosporin A decreases the LTB_4 content in psoriatic lesions. Another possible mechanism has been demonstrated by Hamasaki et al. [71]. They showed that cyclosporin A inhibits leukotriene production in intact rat basophilic leukemia-1 cells by a modulatory effect on one of the intracellular events that includes the activation of the 5-LO [71].

12-HETE is present in high concentrations in chronic psoriatic plaques [72, 73], predominantly in scales [55]. Topical application or intradermal infusion of racemic 12-HETE in normal human volunteers results in an indurated erythematous skin reaction characterized by invasion of both neutrophils and mononuclear leukocytes [30]. Thus, the proinflammatory effect of 12-HETE is much less pronounced than that of LTB_4. It is, therefore, questionable to what extent 12-HETE is involved in the pathogenesis of psoriasis. A potential key role of 12-HETE comes from the fact that normal epidermis itself has very active 12-LO. Therefore, 12-HETE should have its major role as an epidermis-derived mediator of the early events of psoriasis. However, analysis of 12-HETE levels in acute guttate and chronic plaque lesions from the same patient revealed higher 12-HETE levels in chronic plaque lesions than in the acute guttate lesions, and most acute guttate lesions did not contain 12-HETE concentrations sufficient to induce neutrophil chemotaxis [55]. It is therefore likely that 12-HETE is not one of the primary mediators of inflammation in psoriasis.

Analysis of lesional psoriatic skin has shown that 15-HETE is present in both chronic plaques [73, 74] and in acute guttate lesions [55]. In many chronic plaques 15-HETE is present at concentrations capable of inhibiting both the formation and the chemotactic effect of LTB_4 [38, 40]. These data are compatible with the idea that 15-HETE may be important in regulating the magnitude of the inflammatory events in chronic plaque psoriasis.

15-HETE detected in psoriatic skin may originate not only from epidermis and infiltrating neutrophils but also from dermis. Normal human dermis has a larger capacity than normal human epidermis to transform AA into 15-HETE [75]. Furthermore, dermis from uninvolved psoriatic skin generates less 15-HETE than normal dermis *in vitro* [76]. This abnormality may result in defective passage of dermal 15-HETE into epidermis. In support of the idea of the antipsoriatic effect of 15-HETE, repeated injections of 15-HETE caused both clinical and histopathological regression, and sometimes a complete resolution of psoriasis [77]. How 15-HETE exerts its antipsoriatic effect is unknown. *In vitro* data have shown that 15-HETE as well the 15-LO product of dihomo-γ-linolenic acid (DGLA) 15-HETrE are incorporated in the phospholipid membrane preferentially in the phosphatidyl-inositol (PI) fraction in the *sn*-2 position [78, 79]. *In vivo* data have not confirmed the preference for the PI fraction of 15-LO products, although it has been demonstrated that 15-HETE and the 15-LO product of linoleic acid (LA), 13-HODE, are esterified into the phospholipid membrane of normal epidermis [80]. Interestingly, 12-HETE was not detected in these phospholipids, although unesterified 12-HETE was detectable in amounts similar to unesterified 15-HETE [80]. The distribution of monohydroxy fatty acids esterified to phospholipids has also been investigated in lesional psoriatic skin [81]. In accordance with the results obtained in normal skin, 15-HETE and 13-HODE were also found in phospholipids of lesional psoriatic skin, and no esterified 12-HETE was detected. Interestingly, significantly lower levels of 15-HETE and 13-HODE were determined in phospholipids in lesional psoriatic skin compared with matched nonlesional skin [81]. It is therefore possible that part of the antipsoriatic effect of 15-HETE is mediated via incorporation into specific phospholipids. 15-HETE and 13-HODE may also be present in diacyl-glycerol (DAG) by incorporation into the PI fraction of the phospholipids and may thereby modulate DAG-induced activation of protein kinase C (PKC). PKC has been shown to play a key role in the regulation of cell proliferation and differentiation [82]. This hypothesis is underscored by some recent reports from Cho et al. [83–85]. 13-HODE was demonstrated to be incorporated into epidermal phospholipids released by phospholipase C (PLC) activity into 13-HODE containing DAG [83]. This 13-HODE-DAG complex exerts selective inhibition of PKC-b [84], and in a guinea pig model replenishment of hyperproliferative epidermis with topical 13-HODE resulted in accumulation of tissue 13-HODE-DAG and selective inhibition of PKC-b activity [85]. It is therefore possible that part of the antipsoriatic effect of 15-LO products is due to modulation of the DAG activity on various PKC isoforms in the epidermis.

While there appears to be a general increase in the level of LO products in psoriatic lesions, PGE_2 levels are reported to be only modestly elevated [52, 72], and PGE_2 is not believed to play a significant role in the pathogenesis of psoriasis.

Atopic dermatitis

Atopic dermatitis (AD) is a common pruritic and chronic inflammatory skin disease characterized by dryness, erythema and scaling. The dermatitis is symmetrically located preferentially in the face, neck and the flexor aspects of the extremities. An association is often seen with asthma or hay fever. The histopathological changes in the epidermis consist of acanthosis with varying degrees of spongiosis. In dermis a cellular infiltrate is present consisting of lymphocytes, monocytes and mast cells.

The molecular changes leading to the clinical manifestations of AD are unknown. Histamine is one of the mediators suspected to play an important role in the disease processes [86]. Even though histamine is proinflammatory in human skin, its role in AD is uncertain, because of the failure of nonsedative antihistamines to improve the disease. Thus, in the search for potential mediators involved in the inflammatory processes of AD, other mediators than histamine have to be considered. In this respect it would be of great relevance to examine the potential role of AA metabolites in the molecular pathology of AD.

Compared with psoriasis, in which AA is preferentially transformed by LOs, AD appears to be characterized by transformation of AA by both CO and 5-LO. Ruzicka et al. [52] reported elevated LTB_4 levels, but normal PGE_2 levels, in suction blister fluid obtained from patients with AD. In keratome biopsies from patients with AD, LTB_4 and PGE_2 have been determined in biologically active amounts in both lesional and perilesional skin [87]. In contrast, normal levels of eicosanoids were found in the uninvolved skin of these patients [87].

PGE_2 has the capacity to induce wheal and flare when injected into human skin [16] and to modulate the inflammatory responses elicited by other mediators [33, 88]. Both increased vascular permeability and the wheal and flare reactions can be modified when PGE_2 is administered concomitantly with LTB_4. The amounts of both LTB_4 and PGE_2 detected in lesional atopic skin [87] are comparable with the doses used by Acher et al. [16] to induce cutaneous inflammation in human skin.

Enhanced LTB_4 and LTC_4 synthesis in leukocytes from patients with AD has also been reported [89], and the spontaneous release of LTB_4 and LTC_4 from neutrophils has been found to be three times higher in patients with AD than in normal controls [90, 91]. In accordance with these results significantly increased LTA_4 hydrolase activity has been demonstrated in polymorphonuclear leukocytes (PMNs) and in peripheral blood mononuclear cells from patients with severe AD compared with cells from patients with moderate and mild AD and cells from normal controls [92].

Furthermore, a reduction in LTA_4 hydrolase activity in the PMNs from patients with moderate and severe AD was seen after clinically improvement [92]. Taken together, these data strongly support the idea that leukotrienes are intimately associated with the inflammatory processes of AD and that leukotriene-induced processes may be amplified by the presence of PGE_2.

References

1   Ford-Hutchinson AW, Bray MA, Doig MV, Shipley ME, Smith MJ (1980) Leukotriene B, a potent chemokinetic and aggregating substance released from polymorphonuclear leukocytes. *Nature* 286: 264–265

2   Showell HJ, Naccache PH, Borgeat P, Picard S, Valerand P, Becker EL, Sha'afi RI (1982) Characterization of the secretory activity of LTB_4 toward rabbit neutrophils. *J Immunol* 128: 811–816

3   Molski TFP, Nacchache PH, Borgeat P, Sha'afi RI (1981) Simularities in the mechanisms by which formylmethionyl-leucyl-phenylalanine. arachidonic acid and leukotriene B_4 increase calcium sodium influxes in rabbit neutrophils. *Biochem Biophys Res Commun* 103: 227–232

4   Bray MA, Cunningham FM, Ford-Hutchinson AW, Smith MJH (1981) Leukotriene B_4: a mediator of vascular permeability. *Br J Pharmac* 72: 483–486

5   Goldman DW, Goetzl EJ (1982) Specific binding of leukotriene B_4 to receptors on human polymorphonuclear leukocytes. *J Immunol* 129: 1600–1604

6   Kragballe K, Desjarlais L, Voorhees JJ (1985) Leukotrienes B_4, C_4 and D_4 stimulate DNA synthesis in cultured human epidermal keratinocytes. *Br J Dermatol* 113: 43–52

7   Tomita Y, Maeda K, Tagami H (1992) Melanocyte-stimulating properties of arachidonic acid metabolites: possible role in postinflammatory pigmentation. *Pigment Cell Res* 5: 357–361

8   Morelli JG, Hake SS, Murphy RC, Norris DA (1992) Leukotriene B_4-induced human melanocyte pigmentation and leukotriene C_4-induced human melanocyte growth are inhibited by different isoquinolinesulfonamides. *J Invest Dermatol* 98: 55–58

9   Hoffstein ST, Manzi RM, Razgaitis KA, Bender PE, Gleason J (1986) Structural requirements for chemotactic activity of leukotriene B_4 (LTB_4). *Prostaglandins* 31: 205–215

10 Ford-Hutchinson AW, Bray MA, Cunningham FM, Davidson EM, Smith MJH (1981) Isomers of leukotriene B_4 possess different biological potencies. *Prostaglandins* 21: 143–152

11 Ford-Hutchinson AW, Rackham A, Zamboni R, Rokach J, Roy S (1983) Comparative biological activities of synthetic leukotriene B_4 and its omega-oxidation products. *Prostaglandins* 25: 29–37

12 Czarnetzki BM, Mertensheimer R (1985) *In vitro* and *in vivo* chemotaxis of guinea pig

leukocytes toward leukotriene B_4 and its omega-oxidation products. *Prostaglandins* 30: 5–11

13 Camp RDR, Russell Jones R, Brain SD, Woollard PM, Greaves MW (1984) Production of intraepidermal microabscesses by topical application of leukotriene B_4. *J Invest Dermatol* 82: 202–204

14 Paulissen M, Peereboom-Stegeman JHJ, van de Kerkhof PCM (1990) An ultrastructural study of transcutaneous migration of polymorphonuclear leukocytes following application of leukotriene B_4. *Skin Pharmacol* 3: 236–247

15 Soter NA, Lewis RA, Corey EJ, Austen KF (1983) Local effects of synthetic leukotrienes (LTC_4, LTD_4, LTE_4 and LTB_4) in human skin. *J Invest Dermatol* 80: 115–119

16 Archer CB, Page CP, Juhlin L, Morley J, MacDonald DM (1987) Delayed-onset synergism between leukotriene B_4 and prostaglandin E_2 in human skin. *Prostaglandins* 33: 799–805

17 Goetzl EJ, Boeynaens JM, Oates JA, Hubbard WC (1981) Stimulus specificity of the chemotactic deactivation of human neutrophils by lipoxygenase products of arachidonic acid *Prostaglandins* 22: 279–288

18 Wong E, Camp RD, Greaves MW (1985) The response of normal and psoriatic skin to single and multiple topical application of leukotriene B_4. *J Invest Dermatol* 84: 421–423

19 Bauer FW, van de Kerkhof PCM, Maassen-de Grood RM (1986) Epidermal hyperproliferation following the induction of microabscesses by leukotriene B_4. *Br J Dermatol* 114: 409–412

20 Reusch MK, Wastek GJ (1989) Human keratinocytes *in vitro* have receptors for leukotriene B_4. *Acta Derm Venereol* 69: 429–431

21 Dahlen SE, Bjørk J, Hedqvist P, Arfors KE, Hammarstrom S, Lindgren JA, Samuelsson B (1981) Leukotrienes promote plasma leakage and leukocyte adhesion in postcapillary venules: *in vivo* effects with relevance to the acute inflammatory response. *Proc Natl Acad Sci USA* 78: 3887–3891

22 Bisgaard H, Kristensen J, Søndergaard J (1982) The effect of leukotrienes C_4 and D4 on cutaneous blood flow in humans. *Prostaglandins* 23: 797–800

23 Camp RDR, Coutts AA, Greaves MW, Kay AB, Walport M (1983) Responses of human skin to intradermal injection of leukotrienes C_4, D_4 and B_4. *Br J Pharmacol* 80: 497–502

24 Bisgaard H, Ford-Hutchinson AW, Charleson S, Taudorf E (1985) Production of leukotrienes in human skin and conjunctival mucosa after specific allergen challenge. *Allergy* 40: 417–423

25 Shaw RJ, Fitzharris P, Cromwell O, Wardlaw AJ, Kay AB (1985) Allergen-induced release of sulphidopeptide leukotrienes (SRS-A) and LTB_4 in allergic rhinitis. *Allergy* 40: 1–6

26 Wollard PM (1986) Stereochemical difference between 12-hydroxy-5,8,10,14-eicosatetraenoic acid (12-HETE) in platelets and psoriatic lesions. *Biochem Biophys Res Commun* 136: 169–176

27 Wollard PM, Murphy GM, Cunningham FM, Camp R, Greaves MW (1988) Proin-

flammatory effects of 12(R)-hydroxy-5,8,10,14-eicosatetraenoic acid in human skin. *Br J Dermatol* 118: 277

28 Kragballe K, Fallon JD (1986) Stimulated platelet aggregation and abnormal arachidonic acid metabolism in psoriasis. *Arch Dermatol Res* 278: 449–453

29 Cunningham IM, Wollard PM (1987) 12(R)-hydroxy-5, 8, 10, 14-eicosatetraenoic acid is a chemoattractant for polymorphonuclear leukocytes *in vitro*. *Prostaglandins* 34: 71–78

30 Dowd PM, Kobza Black A, Woollard PM, Greaves MW (1987) Cutaneous responses to 12-hydroxy-5, 8, 10, 14-eicosatetraenoic acid (12-HETE) and 5, 12-dihydroxyeicosatetraenoic acid (leukotriene B$_4$) in psoriasis and normal human skin. *Arch Dermatol Res* 279: 427–434

31 Flower RJ, Harwey EA, Kingston WP (1976) Inflammatory effects of prostaglandin E2 in rat and human skin. *Br J Pharmacol* 56: 229–233

32 Crunkhorn P, Willis AL (1971) Cutaneous reactions to intradermal prostaglandins. *Br J Pharmacol* 41: 49–56

33 Basran GS, Morley J, Paul W, Turner-Warwick M (1982) Evidence in man of synergistic interactions between putative mediators of acute inflammation and asthma. *Lancet* 1: 935

34 Lord JT, Ziboh VA (1979) Specific binding of prostaglandin E2 to membrane preparations from human skin: receptor modulation by UVB-irradiation and chemical agents. *J Invest Dermatol* 73: 373–377

35 Goodwin JS, Webb DR (1980) Regulation of the immune response by prostaglandins. *Clin Immunol Immunopathol* 15: 106–122

36 Van Epps DE (1983) Mediators and modulators of human lymphocyte chemotaxis. *Agents Actions-Suppl* 12: 217–233

37 Kunkel SL, Chensue SW, Phan SH (1986) *Prostaglandins* as endogenous mediators of interleukin-1 production. *J Immunol* 136: 186–192

38 Vanderhook JY, Bryant RW, Bailey JM (1980) Inhibition of leukotriene biosynthesis by the leukocyte product 15-hydroxy-5, 8, 11, 13-eicosatetraenoic acid. *J Biol Chem* 255: 10064–10066

39 Chang J, Lamb B, Marinari L, Kreft AF, Lewis AJ (1985) Modulation by hydroxyeicosatetraenoic acids (HETEs) of arachidonic acid metabolism in mouse resident peritoneal macrophages. *Eur J Pharmacol* 107: 215–222

40 Ternowitz T, Fogh K, Kragballe K (1988) Specific inhibition of leukotriene B$_4$-induced neutrophil chemotaxis by 15-hydroxy-eicosatetraenoic acid (15-HETE). *Skin Pharmacol* 1: 93–99

41 Fogh K, Herlin T, Iversen L, Kraballe K (1993) Modulation of eicosanoid formation by lesional skin of psoriasis skin. An ex vivo skin model. *Acta Derm Venereol* 73: 191–193

42 Ternowitz T, Andersen PH, Bjerring P, Fogh K, Schroeder J-M, Kragballe K (1989) 15-Hydroxy-eicosatetraenoic acid (15-HETE) specifically inhibits the LTB$_4$-induced skin response. *Arch Dermatol Res* 281: 401–405

43 Fogh K, Hansen ES, Herlin T, Knudsen V, Henriksen TB, Ewald H, Bünger C, Kragballe

K (1989) 15-Hydroxy-eicosatetraenoic acid (15-HETE) inhibits carragheenan-induced experimental arthritis and reduces synovial fluid leukotriene B_4 (LTB$_4$). *Prostaglandins* 37: 213–228

44   Gualde N, Chable-Rabinovitch H, Motta C, Durand J, Beneytout JL, Rigand M (1983) Hydroperoxyeicosatetraenoic acids: potent inhibitors of lymphocyte responses. *Biochim Biophys Acta* 750: 429–433

45   Gualde N, Atluru D, Goodwin JS (1985) Effect of lipocyclooxygenase metabolites of arachidonic acid on proliferation of human T cells and T cell subsets. *J Immunol* 134: 1125–1129

46   Farrar WL Humes JL (1985) The role of arachidonic acid metabolism in the activities of interleukin 1 and 2. *J Immunol* 135: 1153–1159

47   Conrad DJ, Kuhn H, Mulkins M, Highland E, Sigal E (1992) Specific inflammatory cytokines regulate the expression of human monocyte 15-lipoxygenase. *Proc Natl Acad Sci USA* 89: 217–221

48   Deleuran B, Iversen L, Kristensen M, Field M, Kragballe K, Thestrup-Pedersen K, Stengaard-Pedersen K (1994) Interleukin-8 secretion and 15-lipoxygenase activity in rheumatoid arthritis: *In vitro* anti-inflammatory effects by interleukin-4 and interleukin-10, but not by interleukin-1 receptor antagonist protein. *Br J Rheumatol* 33: 520–525

49   Deleuran B, Iversen L, Deleuran M, Yssel H, Kragballe K, Stengaard-Pedersen K, Thestrup-Pedersen K (1995) Interleukin 13 suppresses cytokine production and stimulates the production of 15-HETE in PBMC. *Cytokine* 7: 319–324

50   Brain S, Camp R, Dowd P, Black AK, Greaves M (1984) The release of leukotriene B_4-like material in biologically active amounts from lesional skin of patients with psoriasis. *J Invest Dermatol* 83: 70–73

51   Grabbe J, Czarnetzki MB, Rosenbach T, Mardin M (1984) Identification af chemotactic lipoxygenase products of arachidonate metabolism in psoriasis. *J Invest Dermatol* 82: 477–479

52   Ruzicka T, Simmet T, Peskar BA, Ring J (1986) Skin levels of arachidonic acid derived inflammatory mediators and histamine in atopic dermatitis and psoriasis. *J Invest Dermatol* 86: 105–108

53   Brain SD, Camp RDR, Kobza Black A, Dowd PM, Greaves MW, Ford-Hutchinson AW, Charleson S (1985) Leukotrienes C_4 and D_4 in psoriatic skin lesions. *Prostaglandins* 29: 611–619

54   Fauler J, Neumann C, Tsikas D, Fröhlich J (1992) Enhanced synthesis of cysteinyl leukotrienes in psoriasis. *J Invest Dermatol* 99: 8–11

55   Fogh K, Herlin T, Kragballe K (1989) Eicosanoids in acute and chronic psoriatic lesions. Leukotriene B_4, but not 12-HETE is present in biologically active amounts in acute guttate lesions. *J Invest Dermatol* 92: 837–841

56   Ford-Hutchinson AW (1993) 5-Lipoxygenase activation in psoriasis: a dead issue? *Skin Pharmacol* 6: 292–297

57   Grabbe J, Rosenbach T, Czarnetzki BM (1985) Production of LTB$_4$-like chemotactic arachidonate metabolites from human keratinocytes. *J Invest Dermatol* 85: 527–530

58 Rosenbach T, Grabbe J, Moller A, Schwanitz HJ, Czarnetski BM (1985) Generation of leukotrienes from normal epidermis and their demonstration in cutaneous disease. *Br J Dermatol* 113 (suppl 28): 157–167

59 Ziboh VA, Casebolt TL, Marcelo CL, Voorhees JJ (1984) Biosynthesis of lipoxygenase products by enzyme preparations from normal and psoriatic skin. *J Invest Dermatol* 83: 248–251

60 Iversen L, Fogh K, Ziboh VA, Kristensen P, Schmedes A, Kragballe K (1993) Leukotriene B_4 formation during human neutrophil keratinocyte interactions: evidence for transformation of leukotriene A_4 by putative keratinocyte leukotriene A_4 hydrolase. *J Invest Dermatol* 100: 293–298

61 Iversen L, Kristensen P, Grøn B, Ziboh VA, Kragballe K (1994) Human epidermis transforms exogenous leukotriene A4 into peptide leukotrienes: possible role in transcellular metabolism. *Arch Dermatol Res* 286: 261–267

62 Solá J, Godessart N, Vila L, Puig L, de Moragas JM (1992) Epidermal cell-polymorphonuclear leukocyte cooperation in the formation of leukotriene B_4 by transcellular biosynthesis. *J Invest Dermatol* 98: 333–339

63 Iversen L, Ziboh VA, Shimizu T, Ohishi N, Rådmark O, Wetterholm A, Kragballe K (1994) Identification and subcellular localization of leukotriene A_4-hydrolase activity in human epidermis. *J Dermatol Sci* 7: 191–201

64 Iversen L, Deleuran B, Hoberg AM, Kragballe K (1996) LTA_4 hydrolase in human skin: Decreased activity, but normal concentration in lesional psoriatic skin. Evidence for different LTA_4 hydrolase activity in human lymphocytes and human skin. *Arch Dermatol Res* 288: 217–224

65 Ikai K, Okano H, Horiguchi Y, Sakamoto Y (1994) Leukotriene A_4 hydrolase in human skin. *J Invest Dermatol* 102: 253–257

66 Iversen L, Kristensen P, Nissen JB, Merrick WC, Kragballe K (1995) Purification and characterization of leukotriene A_4 hydrolase from human epidermis. *FEBS Lett* 358: 316–322

67 McGee J, Fitzpatrick F (1985) Enzymatic hydration of leukotriene A_4. Purification and characterization of a novel epoxide hydrolase from human erythrocytes. *J Biol Chem* 260: 12832–12837

68 Iversen L, Kragballe K, Ziboh VA (1997) Significance of leukotriene A_4 hydrolase in the pathogenesis of psoriasis. *Skin Pharmacol* 10: 169–177

69 Ellis CN, Gorsulowsky DC, Hamilton TA, Billings JK, Brown MD, Headington JT, Cooper KD, Baadsgaard O, Duell EA, Annesley TM et al (1986) Cyclosporine improves psoriasis in a double-blind study. *JAMA* 256: 3110–3116

70 Iversen L, Svendsen M, Kragballe K (1996) Cyclosporin A down-regulates the LTA_4 hydrolase level in human keratinocyte cultures. *Acta Derm Venereol* 76: 424–428

71 Hamasaki Y, Matsumoto S, Kobayashi I, Zaitu M, Muro E, Ichimaru T, Miyazaki S (1995) Cyclosporin A inhibits leukotriene production in intact RBL-1 cells without inhibiting leukotriene biosynthetic enzymes. *Prostaglandins Leukot Essent Fatty Acids* 52: 365–371

72   Hammarstrom S, Hamberg M, Samuelsson B, Duell EA, Starwiski M, Voorhees JJ
      (1975) Increased concentrations of free arachidonic acid, prostaglandin E_2, F_2 and of
      12-hydroxy-5,8,10,14-eicosatetraenoic acid in epidermis of psoriasis: evidence of per-
      turbed regulation of arachidonic acid levels in psoriasis. *Proc Natl Acad Sci USA* 72:
      5130–5134

73   Camp RDR, Mallet AI, Woollard PM, Brain SD, Kobza Black A, Greaves MW (1983)
      The identification of hydroxy fatty acids in psoriatic skin. *Prostaglandins* 26: 431–447

74   Fogh K, Kiil J, Herlin T, Ternowitz T, Kragballe K (1987) Heterogenous distribution of
      lipoxygenase products in psoriatic skin lesions. *Arch Dermatol Res* 279: 504–511

75   Kragballe K, Desjarlais L, Duell EA, Voorhees JJ (1986) Dermis derived 15-hydroxy-
      eicosatetraenoic acid inhibits epidermal 12-lipoxygenase activity. *J Invest Dermatol* 87:
      494–498

76   Kragballe K, Duell EA, Voorhees JJ (1986) Selective decrease of 15-hydroxy-eicosate-
      traenoic acid (15-HETE) formation in uninvolved psoriatic dermis. *Arch Dermatol* 122:
      877–880

77   Fogh K, Søgaard H, Herlin T, Kragballe K (1988) Improvement of psoriasis vulgaris
      after intralesional injections of 15-hydroxyeicosatetraenoic acid (15-HETE). *J Am Acad
      Dermatol* 18: 279–285

78   Miller CC, Tang W, Cho T, Ziboh VA (1993) Selective incorporation of HETEs into epi-
      dermal phospholipids. In: Bailey JM (ed) *Prostaglandins Leukot Lipoxins PAF*. Plenum
      Press, New York, 409–419

79   Heitmann J, Iversen L, Kragballe K, Ziboh VA (1995) Incorporation of 15-hydroxye-
      icosatrienoic acid in specific phospholipids of cultured human keratinocytes and psori-
      atic plaques. *Exp Dermatol* 4: 74–78

80   Grøn B, Iversen L, Ziboh VA, Kragballe K (1993) Distribution of monohydroxy fatty
      acids in specific human epidermal phospholipids. *Exp Dermatol* 2: 38–44

81   Grøn B, Iversen L, Ziboh VA, Kragballe K (1993) Monohydroxy fatty acids esterified to
      phospholipids are decreased in lesional psoriatic skin. *Arch Dermatol Res* 285: 449–454

82   Inohara S (1992) Studies and perspectives of signal transduction in the skin. *Exp Der-
      matol* 1: 207–220

83   Cho Y, Ziboh VA (1994) Incorporation of 13-hydroxyoctadecadienoic acid (13-HODE)
      into epidermal ceramides and phospholipids: phospholipase C-catalyzed release of novel
      13-HODE-containing diacylglycerol. *J Lipid Res* 35: 255–262

84   Cho Y, Ziboh VA (1994) Expression of protein kinase C isoenzymes in guinea pig epi-
      dermis: selective inhibition of PKC-b activity by 13-hydroxyoctadecadienoic acid-con-
      taining diacylglycerol. *J Lipid Res* 35: 913–921

85   Cho Y, Ziboh VA (1994) 13-Hydroxyoctadecadienoic acid reverses epidermal hyper-
      proliferation via selective inhibition of protein kinase C-b activity. *Biochem Biophys Res
      Commun* 201: 257–265

86   Ring J, Sedlmeier F, von der Helm D, Mayr T, Walz U, Ibel H, Riepel H, Przybilla B,
      Reimann HJ, Dorsch W (1986) Histamine and allergic diseases. In: Ring J, Burg G (eds):
      New trends in allergy, vol 2. Springer-Verlag, Berlin, Heidelberg, New York, 45–77

87   Fogh K, Herlin T, Kragballe K (1989) Eicosanoids in skin of patients with atopic dermatitis. Prostaglandin E_2 and leukotriene B_4 are present in biologically active concentrations. *J Allergy Clin Immunol* 83: 450–455

88   Archer CB, Frolich W, Page CP, Paul W, Morley J, MacDonald DM (1984) Synergistic interactions between prostaglandins and Paf-acether in experimental animals and man. *Prostaglandins* 27: 495–501

89   Sampson AP, Thomas RU, Costello JF (1992) Enhanced leukotriene synthesis in leukocytes of atopic dermatitis and asthmatic subjects. *Br J Clin Pharmacol* 33: 423–430

90   Neuber K, Hilger RA, König W (1991) Interleukin-3, interleukin-8, FMLP and C5a enhance the release of leukotrienes from patients with atopic dermatitis. *Immunology* 73: 83–87

91   Hilger RA, Neuber K, König W (1991) Conversion of leukotriene A_4 hydrolase by neutrophils and platelets from patients with atopic dermatitis. *Immunology* 74: 689–695

92   Okano-Mitani H, Ikai K, Imamura S (1996) Leukotriene A_4 hydrolase in peripheral leukocytes of patients with atopic dermatitis. *Arch Dermatol Res* 288: 168–172

Cutaneous essential fatty acids and hydroxy fatty acids: Modulation of inflammatory and hyperproliferative processes

Vincent A. Ziboh

Department of Dermatology, School of Medicine, University of California, Davis, CA 95616, USA

Introduction

The first indication that dietary fat may be essential for healthy growing animals was presented in 1918 by Aron, who proposed that butter has a nutrient value that cannot be provided by other dietary components [1]. This report suggested that a nutrient is inherent in fat apart from its caloric contribution and that this nutrient maybe related to the presence of lipids. In 1929 Burr and Burr [2] presented the first in a series of papers that a "new deficiency disease can be produced by the rigid exclusion of fat from the diet". They suggested that warm-blooded animals in general cannot biosynthesize appreciable quantities of certain fatty acids. In 1930, both investigators added to their earlier work by presenting evidence that the dietary inclusion of linoleic acid (LA) alone could reverse all deficiency symptoms resulting from a fat-free diet, thus, LA or 18:2n-6[*] was heralded as an "essential fatty acid" (EFA) [3]. This pioneering study contained additional observations which indicated that, besides the visible scaliness of the skin, animals with EFA deficiency also experienced increased water loss through the skin. Thus, in these early studies Burr recognized the two major defects that have been associated with EFA deficiency in cutaneous biology, namely, epidermal hyperproliferation and increased permeability of the skin to water.

Structural forms

Two major families of polyunsaturated fatty acids (PUFAs) are characteristic of the mammalian species (the n-6 and n-3 PUFAs). The n-6 and n-3 PUFAs are defined by

---

[*]   Fatty acids and acyl groups are denoted as 18:2n-6, 18:3n-3 and so on, with the first number presenting the number of carbons in the acyl chain. The number following the colon indicates the number of methylene interrupted cis-double bonds and the letter of "n" for "ω" indicates the number of carbon atoms from the methyl end of the acyl chain to the nearest double bond.

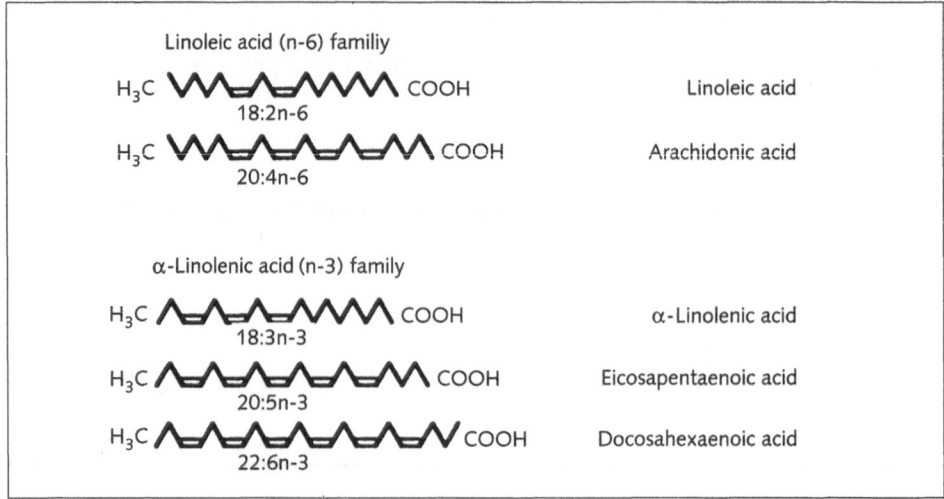

Figure 1
Major families of mamalian polyunsaturated fatty acids.

the position of the double bond closest to the terminal methyl group of the fatty acid molecule. In the n-6 family, for instance, the first double bond occurs between the sixth and seventh carbons from the methyl group end of the molecule, whereas in the n-3 family, the first double bond occurs between the third and fourth carbons. These structural forms are shown in Figure 1. These basic structures cannot be biosynthesized *de novo* in appreciable amounts by vertebrate animals nor are the n-3 and n-6 families of PUFAs interconvertible. Thus, these PUFAs must be obtained from diet.

Dietary sources

The 18-carbon n-6 and n-3 PUFAs are synthesized on land by many plants and therefore are dietarily obtained from vegetable oils. However, the longer chain members of each family are either biosynthesized *in vivo* after dietary ingestion of the shorter 18-carbon precursors or they are obtained directly from animal or marine sources. For example, the longer-chain n-3 PUFAs, eicosapentaenoic acid (EPA, 20:5n-3) and docosahexaenoic acid (DHA, 22:6n-3), are found in fish and marine oils and therefore can be ingested directly from these sources. The important longer chain n-6 PUFA, arachidonic acid (AA, 20:4n-6), is found in liver, brain and meat, which are rich dietary sources of AA.

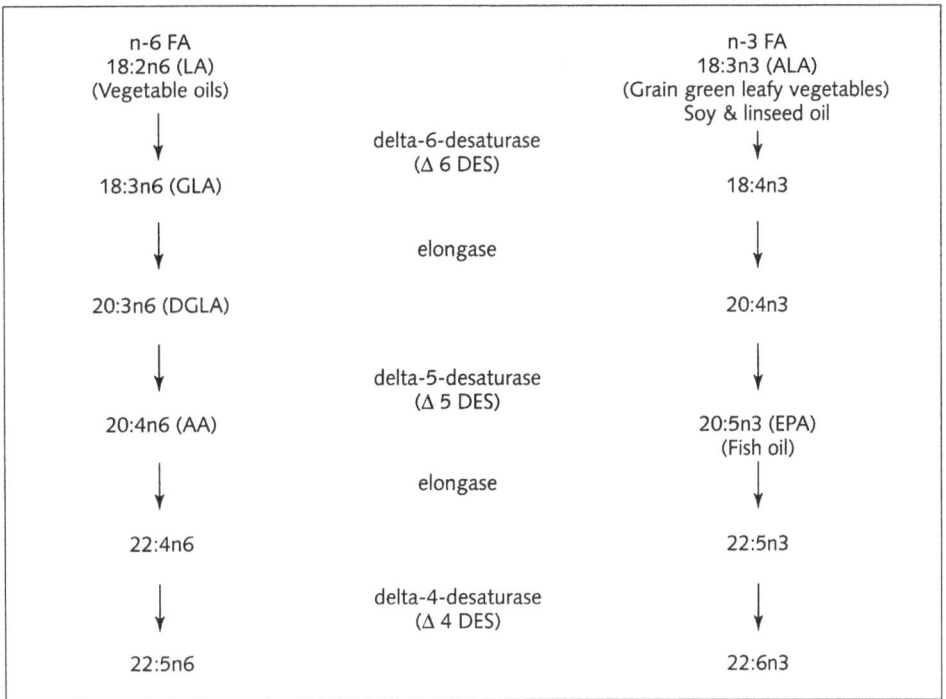

Figure 2
Oxidative desaturation and elongation of n-6 and n-3 polyunsaturated fatty acids.

Metabolism of essential fatty acids

The shorter-chain EFAs, linoleic acid (LA, 18:2n-6) and α-linolenic acid (α-LA, 18:3n-3), serve as the initial unsaturated precursors for *in vivo* biosynthesis of the longer-chain PUFAs. Metabolism of the EFAs in most tissues involves an alternating sequence of Δ^6-desaturation, chain elongation and Δ^5-desaturation in which two hydrogen atoms are removed to create a new double bond followed by the addition of two carbon atoms to lengthen the fatty acid chain [4] (Fig. 2). The desaturations are catalyzed by two separate enzymes: the Δ^6-desaturase catalyzing the transformation of 18:2n-6 (LA) to 18:3n-6 (GLA) and the Δ^5-desaturase catalyzing the transformation of dihomo-γ-linolenic acid (20:3n-6, DGLA) to arachidonic acid (20:4n-6, AA). The elongase enzyme catalyzes the elongation of GLA to DGLA [5]. It is believed that the same enzymes catalyze equivalent steps in the n-3 and n-9 pathways [6]. The PUFA families interact in such a manner that n-3 PUFAs competitively suppress the bioconversion of n-6 PUFAs. Both n-6 and n-3

57

PUFAs respectively do suppress the formation of nonessential long-chain n-9 acids, hence the negligible formation of long-chain n-9 PUFA (20:3n-9) in normally fed animals.

Cutaneous significance of 18-carbon polyunsaturated fatty acids

Linoleic acid (18:2n-6)

Role in skin water barrier system
The most abundant PUFAs in human skin are the 18-carbon linoleic acid (LA) and the 20-carbon arachidonic acid (AA) [7, 8].

One functional significance of the shorter 18-carbon LA is its involvement in the maintenance of the epidermal water barrier [9–12], which is one of the major abnormalities of cutaneous EFA deficiency. The physical structure of the epidermal water barrier is sheets of stacked lipid bilayers, or lamellae, which fill the intercellular spaces of the uppermost layer of the skin epidermis (stratum corneum). These lipid bilayers contain large amounts of sphingolipids [13], of which the linoleate-rich species have been characterized as acylglucosyceramide, acylceramide and a unique unidentified acylacid [14–17].

Role in epidermal hyperproliferation
Although the dietary feeding of LA to EFA-deficient animals is known to reverse the major cutaneous symptoms of EFA deficiency, which include hyperproliferation and increased transepidermal water loss, the mechanism for such reversals remains unclear. The skin is unique among most other tissues in that it lacks the capability to metabolize 18-carbon LA into 20-carbon AA [18]. Interestingly, LA is metabolized by epidermal 15-lipoxygenase into predominantly 13-hydroxy-9,11-octadecadienoic acid (13-HODE), a major hydroxy fatty acid in normal epidermis (Fig. 3).

Functional role of 13-HODE in the epidermis
In vitro incubation of LA with 15-lipoxygenase prepared from human epidermis yields 13-HODE [19]. Topical application of 13-HODE to a guinea pig model of DHA-induced epidermal hyperproliferation resulted in the reversal of the hyperproliferation to a normal state [20]. To delineate a possible mechanism for the effect of topical 13-HODE, we demonstrated that topical 13-HODE was incorporated first into the inositol phospholipids followed by phospholipase C-catalyzed release into 13-HODE-containing diacylglycerol (13-HODE-DAG) [21]. This novel 13-HODE-substituted-DAG was markedly depleted in the hyperproliferative skin of DHA-treated animals, paralleling the increased activities of epidermal protein

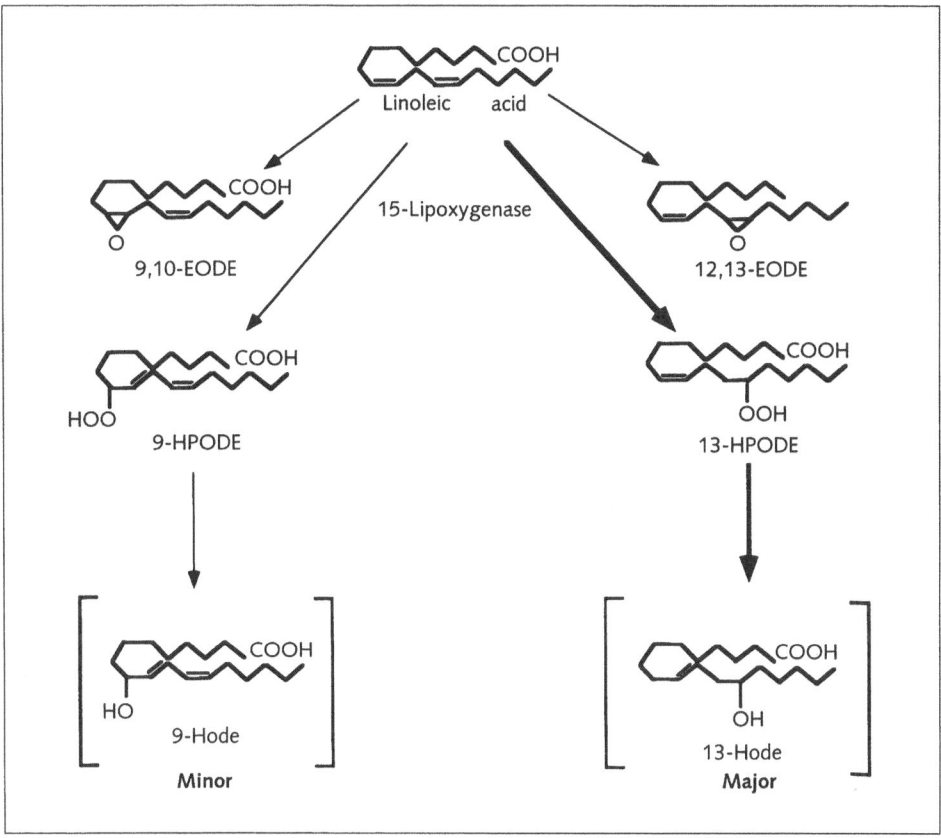

Figure 3
Oxidative metabolism of linoleic acid by human epidermis.

kinase C-α (PKC-α) and PKC-β [22]. Topical replenishment of the hyperprolifera-tive epidermis with topical 13-HODE resulted in accumulation of tissue 13-HODE-DAG, selective suppression of PKC-β activity [13] and reversal of the epidermal hyperproliferation to normal [22].

To establish the *in vivo* relevance of the *in vitro* observations, guinea pigs were made essential fatty acid-deficient (EFAD). Histological examination of the epider-mis of EFAD animals after dietary supplement with coconut oil (which contained negligible LA) demonstrated characteristic epidermal thickening (acanthosis) and hyperplasia (Fig. 4) when compared with epidermis from control animals which were fed safflower oil (which contained LA). In the experimental design, feeding the EFAD guinea pigs a diet supplemented with safflower oil for 2 weeks resulted in dis-

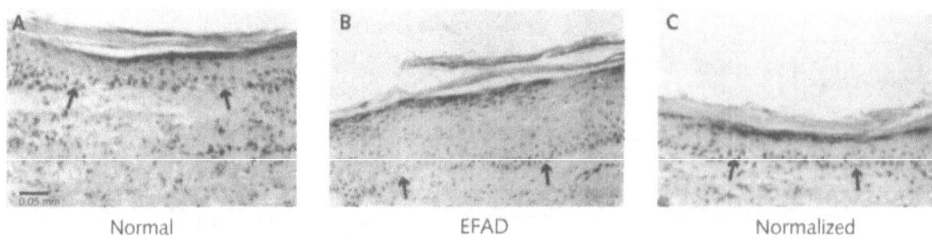

Normal   EFAD   Normalized

Figure 4
Histological evaluation of epidermal slices from control, essential fatty acid-deficient and reversed guinea pigs with H-E (magnification × 200). Comparisons of epidermis from (A) control group, (B) EFAD group and (C) reversed group. The arrows indicate the degree of acanthosis (epidermal thickening) and hyperkeratosis in epidermis of the EFAD group.

appearance of the characteristic acanthosis and hyperplasia. To delineate a mechanism for this reversal, tissue levels of putative 13-HODE-DAG, PKC-isozymes and tissue hyperproliferation were determined during the development and reversal of EFAD. Findings from these experiments revealed that cutaneous 13-HODE and novel 13-HODE-DAG were more markedly reduced in EFAD animals than in the LA-enriched safflower oil-fed control animals. These reductions of 13-HODE and 13-HODE-DAG paralleled both the elevated EFAD-induced epidermal hyperproliferation and the elevated expressions and activities of PKC-α and -β isozymes [23]. Refeeding the animals with LA-enriched safflower oil replenished the *in vivo* tissue level of 13-HODE-DAG. This replenishment, paralleled both the selective down-regulation of PKC-β expression and activity and the reversal of hyperproliferation. These results suggest that maintenance of adequate levels of 13-HODE in epidermis plays a role in modulating cutaneous hyperproliferation *in vivo* via the generation of novel 13-HODE-DAG and the selective suppression of epidermal membrane PKC-β isozyme.

Cutaneous significance of 20-carbon polyunsaturated fatty acids: attenuation of proinflammatory LTB$_4$ generation by PUFA hydroxy fatty acids

Arachidonic acid (20:4 n-6)

The 20-carbon arachidonic acid (AA) is the second prominent PUFA in the skin. It constitutes approximately 6–10% of the total fatty acids in the epidermal phospholipids of guinea pigs [1] and approximately 9% in the epidermal phospholipids of human skin [7]. Its functional role in the epidermis depends largely on its generation of biologically potent oxidative metabolites such as prostaglandins and hydroxy fatty acids. When AA is released from epidermal membrane phopholipids by epidermal cytosolic phospholipase A_2 (cPLA$_2$), AA undergoes oxidative transformation via the cyclooxygenase pathway, generating mainly prostaglandins PGE$_2$, PGE$_{2\alpha}$ and PGD$_2$. Although AA is metabolized via the lipoxygenase pathway in the epidermis, the major metabolic pathway involves the 15-lipoxygenase enzyme. 15-Lipoxygenase catalyzes the transformation of AA in the epidermis into mainly 15-hydroxyeicosatetraenoic acid (15-HETE). Functionally, this hydroxy fatty acid metabolite has been reported as an inhibitor of generation of proinflammatory leukoterine B$_2$ (LTB$_2$) from polymorphonuclear cells (PMNs) and basophils [24]. Consistent with this antiinflammatory potential, 15-HETE was reported to improve the symptoms of psoriasis vulgaris after intralesional injections [25].

Dihomo-γ-linolenic acid (20:3n-6)

γ-Linolenic acid (GLA, 18:3n-6) is a metabolite of Δ^6 desaturase of LA which undergoes rapid elongation by an epidermal elongase into dihomo-γ-linolenic acid (DGLA, 20:3n-6). When taken as a nutritional supplement, GLA has been reported to exert clinical efficacy in a variety of diseases, such as the clinical improvement of patients with atopic eczema [26, 27] and the suppression of acute and chronic inflammation [28–30]. Dietary GLA and its elongation metabolite DGLA have been reported to suppress acute and chronic inflammation and joint tissue injury in experimental animal models [31]. A notable feature of dietary GLA in the epidermis is, first, its *in vivo* elongation to DGLA, followed by its oxidative metabolism via the cyclooxygenase pathway into prostaglandin of the 1-series (PGE$_1$) and its metabolism by 15-lipoxygenase enzyme into 15-hydroxyeicosatrienoic acid (15-HETrE). This possibility was demonstrated after dietary supplementation of guinea pig diet with primrose oil or borage oil (both containing GLA), which resulted in *in vivo* elongation of GLA into DGLA and metabolism into 15-HETrE in the epidermis [32, 33]. Similarly, supplementation of human diet with borage oil (containing GLA) has been shown to elevate *in vivo* levels of DGLA in PMNs. This elevation paralleled the suppression of LTB$_2$ generation from AA by *ex vivo*-activated human PMNs [34]. *In vitro*, 15-HETrE has been shown to markedly inhibit LTB$_2$ generation from AA by rat basophilic leukemia (RBL-I) cells [33]. These *in vitro* effects of 15-HETrE on the inhibition of proinflammatory LTB$_4$ by PMNs seem consistent with the reported beneficial effects of dietary oil containing GLA on inflammatory conditions. To delineate a possible mechanism of action for this hydroxy fatty acid, we recently demonstrated that 15-HETrE was incorporated into the epidermal phosphatidylinositol 4,5-bisphosphate (Ptd Ins 4,5-P$_2$) followed by a subsequate phospholipase C-catalyzed hydrolysis into 15-HETrE- substituted diacylglycerol (15-HETrE-DAG). This finding suggests a modulatory mechanism similar to 13-HODE-DAG.

Eicosapentaenoic acid (20:5n-3) and docosahexaenoic acid (22:6n-3)

Eicosapentaenoic acid (20:5n-3, EPA) and docosahexaenoic acid (22:6n-3, DHA) are the two major PUFAs derived from fish oil. *In vitro* incubations of EPA and DHA with epidermal strips result in the formation of two 15-lipoxygenase products: 15-hydroxyeicosapentaenoic acid (15-HEPE) and 17-hydroxydocosahexaenoic acid (17-HoDHE), respectively [35]. When guinea pig diet is supplemented with fish oil (containing EPA and DHA), the *in vivo* epidermal levels of 15-HEPE and 17-HoDHE are elevated, indicating that both hydroxy fatty acids can be generated from dietary EPA and DHA and possibly exert antiinflammatory effects on cutaneous inflammations.

Possible attenuation of *in vivo* cutaneous inflammatory reactions by hydroxy fatty acids

Since the precursor of the variety of eicosanoids are derived from various dietary EFA sources, it is reasonable to expect that the *in vivo* synthesis of the proinflammatory mediators generated from AA can be modulated by manipulation of dietary PUFAs. The dietary supplementation of diets of psoriatic patients with fish oil containing EPA and DHA have been reported to alleviate the lesions of psoriasis with moderate to excellent results [36–38]. Although the mechanism of this beneficial effect has not been clearly delineated, it is likely that the cutaneous hydroxy fatty acids generated from fish oil EPA and DHA may have exerted some of the beneficial effects, at least in part. This dietary manipulation thus provides an alternative or adjunct protocol for the management of psoriasis or other hyperproliferating epidermal conditions. Such a protocol can be expected to exert negligible side effects. The efficacy of these dietary PUFAs is compatible with *in vivo* results in guinea pigs which indicate that diets supplemented with fish oil containing EPA and DHA do alter the profile of endogenous epidermal phospholipid fatty acids by elevating the levels of EPA and DHA in epidermal phospholipids. After hydrolytic release of EPA and DHA from phospholipids, the PUFAs are presumably metabolized into 15-HEPE and 17-HoDHE, respectively. These metabolites may attenuate the *in vivo* biosynthesis of proinflammatory eicosanoids from AA. Comparative *in vitro* inhibitory effects of monohydroxy fatty acids generated from AA and DGLA (n-6 PUFAs) and EPA and DHA (n-3 PUFAs) on the biosynthesis of PMN-derived proinflammatory LTB_4 is shown in Figure 5. Interestingly, 15-HETrE derived from DGLA is most potent at 20 µM.

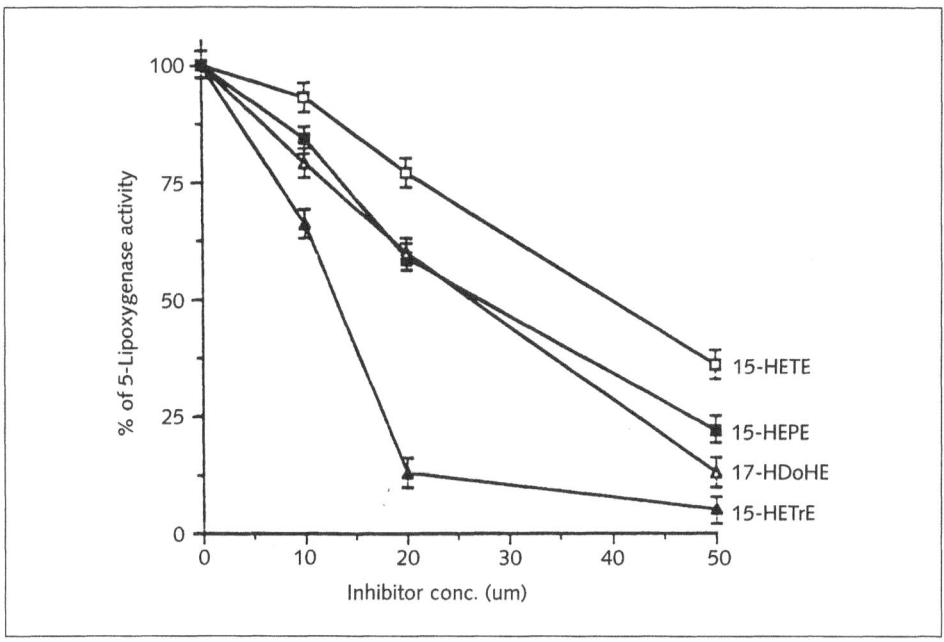

Figure 5
Inhibition of LTB₄ generation from AA by RBL-1 leukemia cells in vitro *by monohydroxy fatty acids: □, 15-HETE; ■, 15-HEPE; △, 17-HDoHE; ▲, 15-HETrE.*

Discussion

Taken together, PUFAs are uniquely metabolized by skin epidermal enzymes, imply-
ing that if present in adequate amounts in the skin epidermis via dietary supplementation, the tissue cyclooxygenase and 15-lipoxygenase enzymes can generate
local *in vivo* antiinflammatory metabolites. In particular, 15-lipoxygenase-catalyzed
metabolites of PUFAs derived from both the n-6 family and the n-3 family do exert
dose-dependent *in vitro* inhibitory effects on the generation of PMN-derived LTB₄
(Fig. 5). Interestingly, 15-HETrE, the metabolite derived from DGLA, exerts the
most potent *in vitro* inhibitory effect when compared with the other monohydroxy
acids derived from AA, EPA and DHA. In contrast, the 18-carbon LA and its oxidative metabolite, 13-HODE, while exerting moderate antiinflammatory effects, has a
potent antiproliferative effect on skin hyperproliferation as evidenced in the reversal of skin hyperproliferation in EFA-deficient animals and DHA-treated skin in
guinea pigs. These findings underscore the significance of 15-lipoxygenation of n-6
and n-3 PUFAs in the skin to generate potent monohydroxylated metabolites which

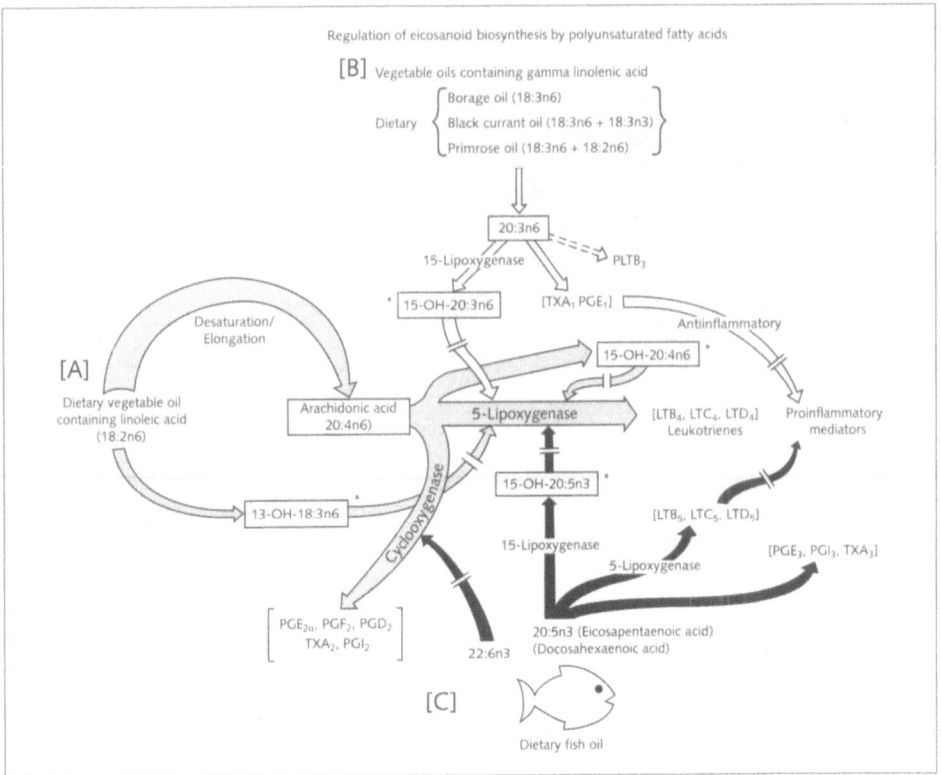

Figure 6
A speculative scenario of the possible modulatory effects of the constituent fatty acids and metabolites from vegetable and fish oils on the generation of proinflammatory leukotrienes from AA. The [A] pathway illustrates the metabolic transformations from dietary intake of safflower oil (rich in LA). It illustrates the in vivo *desaturation and elongation of LA into AA; the 5-lipoxygenation of AA into proinflammatory leukotrienes (particularly leukotriene B₄) by polymorphonuclear cells (PMSs). The [B] pathway illustrates the epidermal transformations of dietary GLA (constituent of evening primrose and borage oils). The GLA undergoes* in vivo *elongation into DGLA, followed by epidermal transformation of DGLA into both 15-HETrE (via the 15-lipoxygenase pathway) and PGE₁ (via the cycloxygenase pathway). The [C] pathway illustrates the metabolic transformations of EPA (constituents of fish oil). EPA is transformed into 15-HEPE and LTB₅.*

seemingly play important roles *in vivo* and function in concert with other cellular processes to attenuate cutaneous inflammatory and hyperproliferative processes.

A speculative scenario of the possible modulatory effects of the constituent fatty acids and metabolites from vegetable and fish oils and their metabolites on the generation of proinflammatory leukotrienes from AA is shown in Figure 6. Thus, it is reasonable to speculate that metabolites from both the B and C pathways could singly or in concert inhibit *in vivo* generation of local tissue proinflammatory leukotrienes (pathway A) from AA in the epidermis. These *in vivo* possibilities imply that the dietary intake of highly purified triglycerides from vegetable or fish oils, or the intake of synthetic structured triglycerides with appropriate constituent PUFAs, may offer an alternative therapeutic modality for alleviating inflammatory disorders.

Acknowledgments
The author thanks Donnelle Yoshino for preparation of the manuscript. Some of the studies referenced in this review were supported in part by Research grant AM30679 from the National Institutes of Health of the United States Public Health Service.

References

1   Aron H (1918) Über den Nährwert. *Biochem Z* 92: 211–233
2   Burr GO, Burr MM (1929) A new deficiency disease produced by the rigid exclusion of fat from the diet. *J Biol Chem* 82: 345–356
3   Burr GO, Burr MM (1930) On the nature of the fatty acids essential in nutrition. *J Biol Chem* 86: 587–621
4   Marcel YL, Christiansen K, Holman RT (1968) The preferred metabolic pathway from linoleic acid to arachidonic acid *in vitro*. *Biochim Biophys Acta* 164: 25–34
5   Fujiwara Y, Okayasu T, Ishibashi T, Imai, Y (1983) Immunological evidence for the enzymatic difference of Δ^6-desaturase from Δ^9- and Δ^5-desaturase in rat liver microsomes. *Biochem Biophys Res Comm* 110: 36–41
6   Brenner RR (1974) The oxidative desaturation of unsaturated fatty acids in animals. *Mol Cell Biochem* 3: 41–52
7   Chapkin RS, Ziboh VA, Marcelo CL, Voorhees JJ (1986) Metabolism of essential fatty acids by human epidermal preparations: evidence of chain elongation. *J Lipid Res* 27: 945–954
8   Yardley HJ, Cummerly R (1981) Lipid composition and metabolism in normal and diseased epidermis. *Pharmacol Ther* 13: 357–383
9   Hansen HS, Jensen B (1985) Essential function of linoleic acid esterified in acylglucosylceramide and acylceramide in maintaining the epidermal water permeability barrier. Evidence from feeding studies with oleate, linoleate, arachidonate, columbinate and a-linoleate. *Biochim Biophys Acta* 834: 357–363
10   Prottey C (1977) Investigation of functions of essential fatty acids in the skin. *Br J Dermatol* 97: 29–38

11 Prottey C, Hartop PJ, Press M (1975) Correction of the cutaneous manifestations of essential fatty acid efficiency in man by application of sunflower-seed oil to the skin. *J Invest Dermatol* 64: 228–234

12 Hartop PJ, Prottey C (1976) Changes in transepidermal water loss and the composition of epidermal lecithin after applications of pure fatty acid triglycerides to the skin of essential fatty acid-deficient rats. *Br J Dermatol* 95: 255–264

13 Elias PM, Brown BE, Fritsch P, Goerke J, Gray GM, White RJ (1979) Localization and composition of lipids in neonatal mouse stratum granulosum and stratum corneum. *J Invest Dermatol* 73: 339–348

14 Wertz PW, Cho ES, Downing DT (1983) Effect of essential fatty acid deficiency on the epidermal sphingolipids of the rat. *Biochim Biophys Acta* 753: 350–355

15 Wertz PW, Downing DT (1983) Ceramides of pig epidermis: structure determination. *J Lipid Res* 24: 759–765

16 Bowser PA, Nugteren DH, White RJ Houtsmuller UMT, Prottey C (1985) Identification, isolation and characterization of epidermal lipids containing linoleic acid. *Biochim Biophys Acta* 834: 419–428

17 Nugteren DH, Christ-Hazelhof E, van der Beek A, Houtsmuller UMT (1985) Metabolism of linoleic acid and other essential fatty acids in the epidermis of the rat. *Biochim Biophys Acta* 834: 419–436

18 Chapkin RS, Ziboh VA (1984) Inability of skin enzyme preparation to biosynthsize arachidonic acid from linoleic acid. *Biochem Biophys Res Comm* 124: 784–792

19 Yamaguchi RY, Miller C, Ziboh VA (1988) Normal epidermis synthesizes 13-hydroxy-9,11-octadecadienoic acid (13-HODE) from linoleic acid. An inhibitor of the 5-lipogenase pathway. *Clin Res* 36, 255A

20 Miller CC, Ziboh VA (1990) Induction of epidermal hyperproliferation by topical n-3 polyunsaturated fatty acids on guinea pig skin linked to decreased levels of 13-hydroxyoctadecadienoic acid (13-HODE). *J Invest Dermatol* 94: 354–358

21 Cho Y, Ziboh VA (1994) Incorporation of 13-hydroxyoctadecadienoic acid (13-HODE) into epidermal ceramides and phospholipids: phospholipase C-catalyzed release of novel 13-HODE-containing diacylglycerol. *J Lipid Res* 35: 255–262

22 Cho Y, Ziboh VA (1994) 13-Hydroxyoctadecadienoic acid reverses epidermal hyperproliferation via selective inhibition of protein kinase C-β activity. *Biochim Biophys Acta* 201: 257–265

23 Cho Y, Ziboh VA (1995) Nutritional modulation of guinea pig skin hyperproliferation by essential fatty acid deficiency is associated with selective down regulation of protein kinase C-β. *J Nutr* 125: 2741–2750

24 Vanderhoek JY, Bryant RV, Bailey JM (1980) Inhibition of leukotriene biosynthesis by the leukocyte product 15-hydroxy-5,5,11,13-eicosatetraenoic acid. *J Biol Chem* 225: 10064–10066

25 Fogh K, Sogaard H, Herlin T, Kragballe K (1988) Improvement of psoriasis vulgaris after intralesional injections of 15-hydroxyeicosatetraenoic acid (15-HETE). *J Am Acad Dermatol* 18: 279–285

26 Lovell CR, Burton JL, Horrobin DF (1981) Treatment of atopic eczema with evening primrose oil. *Lancet* 1: 278

27 Wright S, Burton JL (1982) Oral evening primrose-seed oil improves atopic eczema. *Lancet* 2: 1120–1122

28 Baker DG, Krakauer KA, Tate GA, Laposata M, Zurier RB (1989) Suppression of human synovial cell proliferation by dihomo-γ-linolenic acid. *Arthritis Rheum* 32: 1273–1281

29 Tate G, Mandell BF, Laposata M, Ohliger D, Baker DG, Schumacher HR, Zurier RB (1989) Suppression of acute and chronic inflammation by dietary γ-linolenic acid. *J Rheumatol* 16: 1729–1736

30 Callegari P, and Zurier RB (1991) Botanical lipids: potential role in modulation of immunologic responses and inflammatory reactions. *Rheum Dis Clin N Amer* 17: 415–425

31 Tate G, Mandell BF, Karmali RA, Laposata M, Baker DG, Schumacher HR, Zurier RB (1988) Suppression of monosodium urate crystal-induced acute inflammation by diets enriched with γ-linolenic acid and eicosapentaenoic acid. *Arthritis Rheum* 31: 1543–1551

32 Miller CC, Ziboh VA (1988) Gammalinolenic acid-enriched diet alters cutaneous eicosanoids. *Biochim Biophys Res Commun* 154: 967–974

33 Miller CC, McCready CA, Jones AD, Ziboh VA (1988) Oxidative metabolism of dihomogammalinolenic acid by guinea pig epidermis. Evidence of generation of anti-inflammatory products. *Prostaglandins* 35: 917–938

34 Ziboh VA, Fletcher MP (1992) Dose-response effects of dietary γ-linolenic acid-enriched oils on human polymorphonuclear-neutrophil biosynthesis of leukotriene B$_4$. *J Clin Nutr* 55: 39–45

35 Miller CC, Yamaguchi RY, Ziboh VA (1989) Guinea pig epidermis generates putative anti-inflammatory metabolites from fish oil polyunsaturated fatty acids. *Lipids* 24: 998–1003

36 Ziboh VA, Miller C, Kragballe K, Cohen KA, Ellis CN, Voorhees JJ (1986) Effects of dietary supplementation of fish oil on neutrophil and epidermal fatty acids. *Arch Dermatol* 122: 1277–1282

37 Bittinger SB, Tucker WFG, Cartwright I, Beehen SS (1988) A double-blind, randomized, placebo-controlled trial of fish oil in psoriasis. *Lancet* 1: 378–380

38 Maurice PDL, Allen BR, Barkeley ASJ, Cockbill SR, Bather PC (1987) The effects of dietary supplementation with fish oil in patients with psoriasis. *Br J Dermatol* 117: 599–606

Dietary fatty acids and skin diseases

Rudolf Stadler and Kerstin Schmidt

Department of Dermatology, Medical Centre Minden, Academic Teaching Unit of the University of Münster, Portastraße 7–9, D-32423 Minden, Germany

Introduction

The knowledge of intraepidermal lipid metabolism of keratinocytes and the respective lipometabolic disorders is of fundamental importance for understanding the role of fatty acids in inflammatory skin diseases.

Fatty acids are components of many structures and are involved in various functions within epidermis. They are not only components of the cell membrane, but also play a decisive role in synthesising intercellular substances and are the starting material for eicosanoid metabolism within skin [91]. The whole epidermis, with the exception of the stratum corneum, is able to synthesise lipids; however, during the course of keratinisation the amount and composition of fatty acids changes. While polar lipids, especially phospholipids, predominate the lower layers of the epidermis, there is an increase in neutral lipids (free fatty acids, triglycerides, sterols) and sphingolipids, especially ceramides in the upper epidermal layers [19]. The end product of differentiation of epidermis, the stratum corneum, is described as a two-component model ("brick and mortar model", anuclear corneocytes embedded in lipid-enriched intercellular material) [19, 21] that is responsible for perpetuating the barrier function [24, 20]. Disturbances within this barrier lead to synthesis of free fatty acids and sterols [26, 38].

Essential fatty acids cannot be synthesised by the organism and are therefore essential to the mammal's diet. One of the major essential fatty acids, linoleic acid (C 18:2), is of ω-6 configuration and the basis for prostaglandin synthesis. Linoleic acid is also often a component of ceramide 1, which is believed proposed to play a role in locking together the multiple intercellular lipid membranes in the stratum corneum [106]. Essential fatty acid deficiency leads to impaired barrier function [47] with increased epidermal water loss [72]. Changes within the metabolism of essential fatty acids have complex implications. They are the basis for understanding the aetiology and disorders of inflammatory skin diseases that will be discussed in the following sections.

Fatty Acids and Inflammatory Skin Diseases, edited by J.-M. Schröder
© 1999 Birkhäuser Verlag Basel/Switzerland

Atopic dermatitis

Atopic dermatitis is a multifactoral systemic disease with genetic predisposition. Regarding its pathogenesis, a defect or disturbed maturation of T suppressor cells is postulated. The defect is reflected in disturbed humoral and cellular immunity.

On biochemical analysis, the distinct clinical xerosis is correlated to a clear change in the relative composition of epidermal lipids [47]. The reddening and scaling of the skin correlates with decreased hydration of the stratum corneum and increased transepidermal loss of water, a sign of disturbed barrier function of the skin [73, 103, 104]. In atopic patients this disorder finds its expression in decreased tolerance to lipid solvents, in a disposition for cumulative toxic contact eczema, and in increased bacterial and viral skin infections [73]. Defective water retention and permeability function in atopic dermatitis are due to changes in the composition of epidermal lipids, especially a deficiency of ceramides [50, 52, 74], strikingly ceramide 1 [111].

Ceramide 1 is esterified with linoleic acid and is a very important factor in interlocking the lipid double membrane and the resulting barrier function of the stratum corneum [51]. In atopic patients the skin content of ceramide 1 may be decreased because of reduced activity of β-glucocerebrosidase (reduced synthesis) or increased activity of ceramidase (increased decomposition) [54]. Ceramides of the stratum corneum are the most important lipids responsible for barrier function and the water-binding capacity of the skin [50, 51]. Transepidermal water loss seems to be the regulatory factor for epidermal barrier function: Destruction of epidermal barrier by organic solving agents leads to increased synthesis of fatty acids, cholesterol and sphingolipids, until the composition of epidermal lipids and transepidermal water loss reaches normal conditions again [26, 38, 46]. Animals fed without supplementation of essential fatty acids show a dramatic increase in epidermal water loss [22, 23]. In fatty acid deficiency more ceramides are possibly esterified with oleic acid than with linoleic acid. Oleic acids containing ceramides seem to be unable to connect the lipid bilayer structure resulting in a deficient barrier function of the skin [105, 108].

It is possible that disturbed metabolism of fatty acids in atopic patients plays an important pathogenic role. In 1989 endogenous δ-6-desaturase deficiency was suggested to be key in the aetiopathogenesis of atopy [75]. Linoleic acid is insufficiently metabolised, resulting in deficient formation of γ-linoleic acid and its follow-up products (Fig. 1). The transformation of linoleic acid to γ-linoleic acid and essential components of lipid membrane is reduced [68, 69, 72, 75]. Due to the δ-6-desaturase deficiency, less γ-linoleic acid, dihomo-γ-linoleic acid and arachidonic acid are produced. In atopic patients this deficiency was shown in fatty tissue, monocytes, breast milk [75, 110] and plasma [69]. The production of prostaglandin E_1 from dihomo-γ-linoleic acid and prostaglandin E_2 from arachidonic acid is also impaired [68]. Arachidonic acid synthesis and prostanoid biosynthesis of atopic monocytes

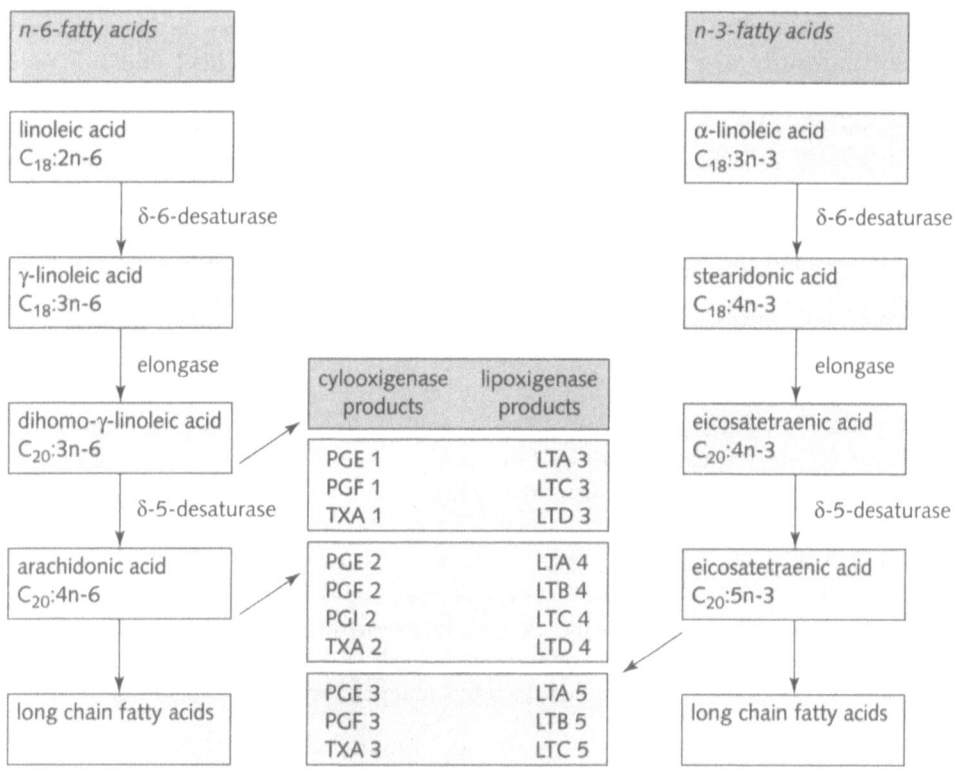

Figure 1
Metabolic pathways of n-3 and n-6 fatty acids [114]

[64] are restricted. As a consequence there is a lower PGE$_2$ production of atopic monocytes compared with normal monocytes [28]. Even in inflammatory skin of atopic patients, no increase in prostaglandin E$_2$ can be observed [85].

E prostaglandins have been reported to play an important role in the regulation of cellular and humoral responses [33]. Regarding cellular immune response, PGE$_1$ supports the maturation of T cells in thymus, especially inducing T cell differentiation [32] and function of T suppressor lymphocytes [81, 102]. In addition, E prostaglandins exert an antiinflammatory effect by inhibiting T lymphocyte activation and proliferation in mitogen-stimulated cells [34, 98], associated with inhibition of interleukin-2 (IL-2) production [101]. In vitro experiments proved that prostaglandin E also impedes the immunoglobulin E (IgE) synthesis of mononuclear blood cells [77].

On the other hand, there is evidence that certain unsaturated fatty acids exert antiinflammatory and immunmodulatory effects themselves [114] and not only through their oxygenation products, the eicosanoids. Dihomo-γ-linoleic acid and arachidonic acid have been reported to inhibit IL-2 production by mitogen-stimulated human T cells and also IL-2 dependent human T cell proliferation by a prostaglandin-independent mechanism [87, 88].

Because of the widespread effects of linoleic/γ-linoleic acid deficiency in atopic patients, several therapeutic approaches for oral or topical substitution have been promoted.

Three double-blind studies analysed the effect of locally applied linoleic/γ-linoleic acid on atopic dermatitis [1, 11, 66] (Tab. 1). In one study significant improvement of atopic dermatitis was verified only by the patients; the doctors assessed no difference. No difference between placebo and verum group was seen in either of the other trials. Measurements of fatty acids revealed a small rise in dihomo-γ-linoleic acid ($p < 0.05$) in plasma of the group treated locally with evening primrose oil [66].

Nevertheless, an open study testing patients with atopic disposition but without clinical symptoms revealed a significant increase in epidermal lipid content and positive influence on epidermal water loss of an emollient containing 12.5% evening primrose oil compared with nontreated areas [53]. In contrast, no effect of a lotion containing 5% essential fatty acids on transepidermal water loss has been reported by another open study [99].

Table 1 - Double-blind studies on local application of linoleic/γ-linoleic acid in atopic dermatitis

Study	Patients	Application	Clinical results	Biochemical results
Macdonald 1985	$n = 20$	emollient containing 10% evening primrose oil vs olive oil	similar rates of healing	verum group: small rise of dihomo-γ-linoleic acid in plasma
Anstey 1990	$n = 12$	evening primrose oil vs placebo	significant increase in patient score, no difference in doctor score	
Borelli 1994	$n = 100$	emollient + 10% borage oil vs emollient	no subjective or objective difference between the two groups	

The question of which concentration of locally applied polyunsaturated fatty acids might be of use in atopic dermatitis remains to be answered [76]. In addition, it is still unknown whether other lipids contained in the base used exert any effect on atopic dermatitis [53]. It may be that a combination of various lipids in the externum used for topical treatment plays an important role [67].

Animal experiments showed that raised transepidermal water loss of rodents fed linoleic acid-free food could be normalised by topical application of linoleic acid and γ-linoleic acid. No normalisation could be reached by applying other unsaturated fatty acids [43, 82]. The restitution of the barrier function and subsequent normalisation of transepidermal water loss was achieved by direct intake of fatty acids in the structural lipids of the epidermis, of which ceramide 1 especially is dependent on linoleic acid [107].

With the idea that patients suffering from atopic dermatitis have a systemic deficiency of δ-6-desaturase, systemic application of γ-linoleic acid might be useful in treatment of this disease. Substituting the deficient products of the δ-6-desaturase metabolism was hypothesised to lead to normalisation of the humorally and cellularly disturbed immune system in atopic patients, especially in children. In atopic patients substitution of linoleic acid is not sufficient to raise the level of linoleic acid metabolites in serum and tissue, as the missing δ-6-desaturase is the key enzyme within the metabolism cascade of ω-6-fatty acids [71]. These findings resulted in the concept of a dietary substitution with γ-linoleic acid with evening primrose oil or borage oil. A possible prevention of atopic dermatitis in newborn babies by supplementing ω-6 fatty acids was one aspect that had to be analysed.

Prolonged breast feeding seems to have a prophylactic effect on the development of atopic disease [86]. However, there are also epidemiologic data that show a higher prevalence of atopy in children having an atopic mother than in children having an atopic father [63].

Early onset of baby food based on cow milk leads to a high prevalence of atopic disease [55]. Compared with the milk of mammals, for instance cow milk, human breast milk contains high concentrations of linoleic acid, γ-linoleic acid and dihomo-γ-linoleic acid [2, 30, 55]. Nevertheless, breast milk of atopic mothers has been reported to contain more linoleic acid and significantly less γ-linoleic acid, dihomo-γ-linoleic acid and arachidonic acid [49]. Similar results were seen in lipid fractions of breast milk from mothers whose children newly developed atopic dermatitis: there were increased proportions of linoleic acid and significantly decreased proportions of δ-6-desaturase dependent metabolites [13]. These results are consistent with findings of abnormal fatty acid status in atopic subjects [13]. A supplementation of maternal diet with preformed long-chain polyunsaturated acids (evening primrose oil, 4 g daily) results in an increase of these dietary fatty acids and total fat content in breast milk [15]. The effects of supplementing the diet of breastfeeding mothers with long-chain polyunsaturated acids on the development of atopic disease in their children further studies require further investigation [13].

Several clinical trials have been carried out to evaluate the effect of orally sub-stituted evening primrose oil containing γ-linoleic acid in atopic eczema.

In a meta-analysis [79] of nine randomised, placebo-controlled, double-blind studies (four parallel, five crossover studies), 311 patients with atopic eczema were evaluated. The medication ranged from 4 to 19 capsules of evening primrose oil (each capsule containing ≈500 mg) daily; active treatment ranged from 4 (one study) to 8 (one study) to 12 (seven studies) weeks. During each study the best avail-able topical treatment was individually continued; 29 patients continued oral med-ication of potent steroids. There were three sets of scores: patient scores, doctor scores (both including measures of inflammation, dryness, scaliness and overall skin involvement) and itch scores. Four studies analysed plasma samples to evaluate the effect of evening primrose oil on lipid levels. The parallel studies revealed a highly significant improvement for evening primrose oil ($p < 0.0001$) in patient and doctor scores. The crossover trials showed the same results without clinical significance. Regarding the itch scores, there was no response to placebo, but all trials except one showed a highly significant response to evening primrose oil. Analyses of plasma phospholipid fractions taken in four of the nine trials (n = 78) revealed a weak but significant correlation (p = 0.022) to clinical scores. Increased arachidonic acid was associated with good clinical response; increased dihomo-γ-linoleic acid was less clearly associated with clinical response. The best clinical response was associated with increased arachidonic acid as well as dihomo-γ-linoleic acid [79].

Significant improvement of atopic dermatitis under supplementation of evening primrose oil was also seen in three other double-blind placebo-controlled studies [7, 69, 90] (Tab. 2), but there are also two double-blind randomised trials and one open study reporting unresponsiveness of atopic eczema to evening primrose oil [3, 6, 94].

One double-blind placebo-controlled crossover trial that included 123 patients did not show any change in clinical symptoms and in the level of fatty acids in plas-ma [3]. As the meta-analysis [79] established a strict efficacy relation to itch, and as other studies showed an increase in dihomo-γ-linoleic acid in serum under evening primrose oil, the lack of change was interpreted as a reduction in patient compliance with taking their medication or a mixup in medication leading to those results [49, 79].

Another double-blind, parallel study showed no improvement of atopic der-matitis after 16 weeks of treatment with 6 g of evening primrose oil daily [6]. A syn-ergistic effect of the addition of n-3 fatty acids to evening primrose oil could not be observed. However, in an intention-to-treat analysis that was carried out later and that included all the patients who had dropped out of the study, found improvement in clinical symptoms [49].

The effect of supplementing diet with evening primrose oil on the clinical course of atopic eczema was also checked in children. In an open study with children suf-fering from atopic eczema (n = 12), improvement of clinical symptoms was observed in administration of 3 g of evening primrose oil daily over 20 weeks [8]. Similar results were obtained in a double-blind parallel study in children suffering from atopic eczema (n = 51) who were given a dose of 0.25 to 0.5 g/kg body weight evening primrose oil daily over 8 weeks. Treatment with evening primrose oil led to an increase of unsaturated fatty acids in plasma, neutrophils, lymphocytes and erythrocytes [7, 8]. This and a further study with 10 children showed no correlation between immune status (measured on IgE) and the response to evening primrose oil [7, 27]. However, significant improvement, especially of itching, was seen in both trials under the study medication (3 g daily or 0.25–0.5 g/kg body weight daily) [7, 27], and because of this improvement a clear reduction in the use of antihistamines was observed [27].

Hyperactivity of children occurring in the course of atopy is variable. In one trial 14 children suffering from atopic eczema showed an improvement in hyperactivity under medication with evening primrose oil [39]. In contrast, a double-blind placebo-controlled study in 58 children with atopic eczema receiving evening primrose oil showed no therapeutic effect regarding the fidgeting [44].

However, it has to be accepted that there are conflicting clinical results and that biochemical data are also not completely consistent. In a meta-analysis, clinical response was associated with an increase in arachidonic acid and dihomo-γ-linolenic acid [79]. Similar results have been obtained by another double-blind study analysing plasma lipid fractions of patients with atopic dermatitis and healthy persons [69].

In the group of patients with atopic eczema there was a significant elevation of linoleic acid in plasma compared with healthy people. All metabolites of linoleic acid were significantly decreased, indicating a δ-6-desaturase deficiency in atopic patients. Oral administration of evening primrose oil lead to a significant increase of the metabolites arachidonic acid and dihomo-γ-linolenic acid. In particular, the increase in dihomo-γ-linolenic acid was dose-dependent (4–6 g of evening primrose oil daily) [69].

A Finnish study also observed an increase in dihomo-γ-linolenic acid in patients suffering from atopic eczema taking evening primrose oil compared with a control group. In contrast, there were no changes in prostaglandin, thromboxane or arachidonic acid levels [90]. Clinical symptoms improved significantly in evening primrose oil-treated patients, which led to a reduction of topical steroids being used [90].

A dose-dependent increase in the content of dihomo-γ-linolenic acid in neutrophils was observed in atopic patients (n = 15) that have been treated with evening primrose oil [89]. Skin biopsies revealed a significant increase in dihomo-γ-linolenic acid in epidermis at highest doses (6 g of evening primrose oil daily) and a significant increase in the ratio between n-6 and monounsaturated fatty acids. In contrast, shave biopsies and acetone-mediated lipid extraction taken from patients (n = 10) with atopic dermatitis revealed a clear decrease in ceramides and free fatty acids in the stratum corneum. But under medication with 6 g of evening primrose oil daily,

Table 2 - Double-blind placebo-controlled studies on evening primrose oil supplementation in atopic dermatitis

Study	Patients	Supplementation	Clinical results	Biochemical results
Manku 1984	adults with atopic eczema (n = 41) healthy people n = 50	2–6 g evening primrose oil daily vs placebo over 12 weeks	dose-dependent (4–6 g/day) significant improvement of eczema	patients with atopic eczema compared with healthy people: elevation of linoleic acid and decrease of its metabolites in plasma. Evening primrose oil leads to an increase in arachidonic acid and dihomo-γ-linoleic acid (dose-dependent, 4–6 g/day)
Bamford 1985	infants (n = 49) and adults (n = 74) with atopic dermatitis	evening primrose oil 2 or 4 g/day (children), 6 or 8 g/day (adults) over 12 weeks	no significant effect on erythema, scale, excoriation, lichenification or overall severity	only increase in plasma arachidonic acid in children taking the low child's dose (n = 6) ⇒ random event
Schalin-Karrila 1986	adults with atopic eczema (n = 25)	evening primrose oil, 4 g/day vs placebo over 12 weeks	significant reduction of in-flammation, severity, dryness itching and body surface involved in treated group	significant rise of dihomo-γ-linoleic acid in plasma in treated group. No treatment effect on blood levels of prostaglandins and thromboxane
Guenther 1987	infants with atopic dermatitis and fidgeting (n = 44)	evening primrose oil, 2–4 g/day age-depen-dent vs placebo over 12 weeks	dermatitis: no difference between treated and placebo groups; improvement in both . groups. Fidgeting: significant decrease in treated group	

Study	Patients	Treatment	Clinical result	Biochemical result
Morse 1989 meta-analysis of nine studies - Burton (children) - Burton (adults) - Gould - Allen - Wexler - Jansen - Holt - Meigel - Bordoni	atopic dermatitis with onset in childhood ($n = 311$)	4–19 capsules (each containing 500 mg of evening primrose oil)/day vs placebo over 4(1), 8 (1) and 12 (7) weeks	itch score: highly significant response to evening primrose oil (eight trials) clinical score: highly significant response to evening primrose oil ($p < 0.0001$) (four trials)	rise of dihomo-γ-linoleic acid and arachidonic acid was weakly significantly correlated with clinical response ($p = 0.022$) (four trials)
Berth-Jones 1993	infants and adults with atopic eczema ($n = 123$)	evening primrose oil, 6 g/day vs evening primrose oil + fish oil, 6 g/day vs placebo over 16 weeks	no improvement	
Biagi 1994	infants with atopic dermatitis ($n = 51$)	evening primrose oil, 0.25 or 0.5 g/kg body weight vs placebo over 8 weeks	significant improvement of clinical symptoms	dose-dependent increase of n-6 fatty acid in erythrocyte cell membrane. Immune status measured by IgE is no predictor for response to evening primrose oil
Hederos 1996	infants with atopic dermatitis ($n = 60$)	evening primrose oil, 4–6 g (age dependent) vs placebo over 16 weeks	dermatitis: no difference between treated and placebo group, improvement in both groups fidgeting: no improvement asthma: no improvement	evening primrose oil group: highly significant increase of dihomo-γ-linoleic acid and arachidonic acid in plasma, no relationship between blood analysis and itching symptoms

no change in these levels was seen despite clinical improvement and decreased transepidermal water loss [48].

A German group [80] did not observe any difference in fatty acid distribution in atopic patients (n = 60) and healthy people (n = 15). However, after taking evening primrose oil, the atopic patients showed a significant rise in dihomo-γ-linoleic acid in blood. The clinical symptoms were clearly improved under a dose of 6 g daily over 4 weeks. As fatty acids are metabolised quickly (maximum results of linoleic acid, γ-linoleic acid, dihomo-γ-linoleic acid and arachidonic acid after 4 to 6 h) the time span between application of medication and taking blood samples is of utmost importance. Many studies did not consider this factor, and this might be an explanation for the partly conflicting results [80].

In conclusion, it has to be stated that substitution with γ-linoleic acid is a well-tolerated additional therapy in atopic eczema and has hardly any side effects. The clinical effect can only be described as mild or restricted. Data obtained from double-blind placebo-controlled studies are inconsistent. Although the majority of the studies found significant improvement of atopic eczema under evening primrose oil, two large studies showed no effect of evening primrose oil on atopic dermatitis. Even when the eczemic reaction responded to orally administered polyunsaturated fatty acids, only a partial reduction of antihistamines could be proven [27]. The application of oral steroids could not be stopped [79]. These studies did not state any reduction of topical steroids.

Taking these facts into consideration, there is no change in the basic therapy regimen of atopic dermatitis by supplementation with evening primrose oil; it seems only to have possible additional effects. The clinical and biochemical results stress that misregulation of fatty acids are relevant in atopic patients; however, they only present one part of the whole aetiopathogenesis of atopic dermatitis.

Psoriasis vulgaris

Psoriasis vulgaris is a multifactionally caused disease whose characteristics are increased growth regulation and cell differentiation of the epidermis as well as inflammatory infiltration of the dermis. Regarding the role of fatty acids in the aetiopathology of psoriasis vulgaris, there is evidence of increased arachidonic acid metabolism in the skin of psoriatic patients [18].

Arachidonic acid is part of the phospholipid fraction of the cellular membrane. It is directly released by the enzyme phospholipase A1 or indirectly by phospholipase C [5]. On the one hand, free arachidonic acid can be transformed by cyclooxygenase to prostaglandins, on the other hand by lipoxygenases to monohydroxy fatty acids and leukotrienes, in particular 5-hydroxyeicosatretanoic acid, leukotriene B and 12-hydroxyeicosatetranoic acid, which all have inflammatory and chemotactic characteristics.

Many results indicate that arachidonic acid seems to play a role in the pathogenesis of psoriasis. In psoriatic plaques an increased concentration of arachidonic acid and 12-hydroxyeicosatetranoic acid [42] and many monohydroxy fatty acids, especially 12-hydroxyeicosatetraenic acid [14], were found. Involved areas from psoriatic patients contained a statistically significant 7- to 11-fold increase in the levels of leukotriene B4 and 15- and 12-hydroxyeicosatetranoic acid in comparison with normal skin [18]. Even uninvolved skin from psoriatic patients contained 40–100% increases in these levels compared with skin from nonpsoriatic subjects [18]. As phospholipase A2 activity was found to be significantly higher in both uninvolved and involved skin from psoriatic patients, phospholipase C activity in contrast was elevated only in lesional epidermis. There is some evidence that phospholipase C especially contributes to arachidonic acid accumulation observed in psoriatic tissue [5].

Leukotriene B4, which has been reported to be one of the most potent chemotactic factors [29], was found in high concentrations in lesional psoriatic skin [12, 18, 35]. Correspondingly, there have been reports regarding increased activity of 5-lipoxygenase in lesional skin compared with uninvolved skin [113]. Epidermal application of leukotriene B4 resulted in infiltration of the dermis by neutrophils and formation of intraepidermal microabscesses [14]. In contrast, leukotriene B4 is not able to produce a complete psoriatic lesion [109]. Therefore, more complex interactions of molecular and cellular mechanisms are likely in which arachidonic acid metabolites may be involved [83]. High leukotriene B4 levels have also been found in skin from patients suffering from other inflammatory diseases as atopic dermatitis [84] and allergic contact dermatitis [4]. Thus, high leukotriene B4 levels in psoriatic skin do not seem to be specific for this disease alone.

One therapeutic approach to psoriasis vulgaris attempts to regulate arachidonic acid metabolism by substituting arachidonic acid with other unsaturated fatty acids [60]. Much interest has been focused on eicosapentaenoic acid, which can act as substrate for lipoxygenase resulting in metabolites having much less biological activity than their eicosanoid counterparts. Therefore, leukotriene B5 derived from eicosapentaenoic acid is about 30 times less potent than leukotriene B4 as a chemotactic agent [31]. High doses of leukotriene B5 have been reported to inhibit the stimulatory effects of leukotriene B4 on keratinocyte DNA synthesis, whereas leukotriene B5 is less potent in stimulating DNA synthesis [61].

Because fish oil is rich in eicosapentaenoic acid and Eskimos (who consume quantities of eicosapentaenoic acid) appear to be free from psoriasis [62], attempts have been made to treat psoriasis with concentrated fish oil [113].

Several placebo-controlled studies have been carried out on the effect of n-3 fatty acid supplementation on psoriasis (Tab. 3). Most of the studies used fish oil containing high amounts of eicosapentaenoic and docosahexaenoic acid. Strikingly, different results were seen: while three groups reported significant improvement of acute guttata psoriasis and stable chronic psoriasis [9, 37, 41], no clinical improve-

Table 3 - Double-blind placebo-controlled studies on fish oil supplementation in psoriasis

Study	Patients	Supplementation	Clinical results	Biochemical results
Veale 1994	psoriatic patients with skin and joint problems vs	evening primrose oil + fish oil vs placebo drug requirement possible	no clinical improvement of skin lesions and arthritis, no reduction of nonsteroid antiinflammatory	leukotriene B4 production during treatment decreased and afterwards increased
Grimminger 1993	acute guttate psoriasis	n-3 fatty acid emulsion infusion vs n-6 lipid emulsion infusion	significant improvement of clinical symptoms ($p < 0.05$)	10-fold increase in neutrophil EPA-derived 5-lipoxygenase product formation
Soyland 1993	moderate to severe psoriasis	fish oil vs corn oil	no significant change in either group	fish oil group: clinical improvement was not correlated with increase of n-3 fatty acids in serum corn oil group: significant correlation between clinical improvement and increase in EPA and total n-3 fatty acids in serum
Gupta 1990	psoriasis oil vs olive oil	fish oil vs olive oil both groups topical steroids	no significant difference between both groups	–
Gupta 1989	stable plaque psoriasis	fish oil + UVB vs placebo + UVB	fish oil group statistically significant greater improvement in all clinical parameters	–
Bittiner 1988	stable chronic psoriasis	fish oil vs olive oil	fish oil group significantly less itching, erythema and scaling	–
Bjorneboe 1988	psoriasis	fish oil vs olive oil	no significant change in clinical manifestation	significant increase in n-3 fatty acids in serum phospholipids

ment of skin symptoms was seen in four other double-blind placebo-controlled trials [10, 40, 95, 100]. Though decreased leukotriene B4 production during n-3 fatty acid supplementation indicated an antiinflammatory effect, psoriatic patients suffering from psoriatic arthritis were not able to reduce their nonsteroidal antiphlogistic requirements during supplementation with fish oil [100].

Open-study testing on fish oil supplementation in psoriasis has revealed conflicting results, mostly showing moderate to mild clinical response [70, 113], and in a few patients excellent to mild improvements [57, 59]. No clinical improvement was seen in plaque-type psoriasis, whereas one patient suffering from generalised pustular psoriasis showed an improvement of his disease [56]. Regarding biochemical analysis, eicosapentaenoic acid (EPA) and docosahexaenoic acid (DCHA) were rapidly incorporated in serum [10, 95, 113], neutrophils and epidermis [113], and high EPA/DCHA ratios correlated with improved clinical response [113]. Nevertheless, clinical improvement in a fish-oil treated group was not correlated with an increase in n-3 fatty acids in serum, but there was a significant correlation between increase in EPA and n-3 fatty acids in serum and clinical improvement in the corn oil (= placebo) group [95]. During fish oil supplementation some groups reported an inhibition of leukotriene B4 production [56, 70] and an increase in neutrophil EPA-derived 5-lipoxygenase product formation [37]. Decreased leukotriene B4 levels did not correlate with clinical therapy [56]. The discrepancy between the high degree of inhibition of leukotriene B4 and the modest clinical effect suggests that leukotriene B4 is not the only mediator involved in the pathogenesis of psoriasis [70].

Regarding the effects of topically applied n-3 fatty acid on psoriatic skin, a multicentre double-blind, placebo-controlled study reported no statistically or clinically relevant difference between ω-3 polyunsaturated fatty acids (10 or 1%) and placebo [45]. Nevertheless, testing fish oil vs. olive oil under occlusive dressings revealed significant improvement in plaque thickness in the fish oil group [25].

In conclusion, supplementation of eicosapentaenoic acid in psoriasis seems to have some antiinflammatory influence, as evidenced by leukotriene B4 reduction. In contrast, clinical symptoms only improved in three of seven double-blind studies, and the clinical value of fish oil supplementation remains uncertain. Fish oil supplementation may exert some positive effects when used in addition to established antipsoriatic therapies, but at present its use as monotherapy is not justifiable.

Acne vulgaris

Acne vulgaris is also a disease caused by many factors and is aetiopathically characterised by sebaceous gland hypertrophy accompanied by increased sebum production, abnormal follicular keratinization and inflammation.

Regarding the role of dietary fatty acids in the aetiopathogenesis of acne, there is evidence of a coherence of sebum linoleate concentration and comedo formation

[16]. Essential fatty acid deficiency leads to sebaceous gland hypertrophy and hyperkeratinization of the ducts, two features relevant to acne [93]. As high rates of sebum production per sebocyte result in low levels of linoleate in the sebaceous esters leading to characteristic hyperkeratosis and comedo formation, local deficiency of linoleic acid was hypothesized to be an aetiologic factor in acne [16, 17, 107]. Linoleic acid content in the ceramide fraction of the stratum corneum has been reported to be much higher than in the ceramide fraction of comedones [107]. Hypothetically, the amount of linoleic acid in a sebaceous cell is constant and decreases only in relation to the increased rate of sebum synthesis [17]; thus, a low concentration of linoleate in sebum is possibly the result of elevated lipogenesis in individual sebaceous cells [17]. In a recent study stratum corneum sphingolipids of patients with acne was reported to be lower than in control subjects [112]. Lower sphingolipid content led to a diminished water barrier function, suggesting that an impaired water barrier function may be responsible for comedo formation, as dysfunction of this barrier is accompanied by hyperkeratosis of the follicular epithelium [112]. The basis for this hypothesis is thus not necessarily a systemic deficiency of linoleic acid [17]. Analyses of plasma phospholipids have revealed a normal content of dihomo-γ-linoleic acid in patients suffering from acne [36].

Androgenic stimulation of glands is known to lead to an increase in lipid synthesis. Use of anabolic-androgenic steroids has been reported to cause an increase in skin surface lipids, especially cholesterol and free fatty acids, in skin biopsy specimens [92]. Androgen action in skin depends on the conversion of testosterone by 5-α-reductase to 5-α-dihydrotestosterone, which then binds to the androgen receptor to regulate specific gene expression [65]. Certain unsaturated fatty acids, such as γ-linoleic acid, have been reported to be potent 5-α-reductase inhibitors, suggesting a link between unsaturated fatty acids and androgen action [65]. Puberty is often accompanied by a rise in acne lesions, possibly because of a local decrease of linoleic acid [17].

As some antiacne agents have been reported to decrease the sebum secretion rate and induce changes in comedonal lipid composition [74, 96, 97], there is also evidence of an influence of sebum synthesis on the concentration of fatty acids, especially linoleic acid in comedonal lipid fraction [49]. However, clinically controlled randomised studies to evaluate the influence of changes in comedonal lipid composition in acne patients and correlating clinical results have not been carried out. Whether oral or topical substitution of essential fatty acids will become necessary in antiacne therapy remains unclear and requires further investigation.

Acknowledgment
We thank A. Licht-Mbalyohere for working with us on this chapter.

References

1   Anstey A, Quigley M, Wilkinson JD (1990) Topical evening primrose oil as treament for atopic eczema. *J Dermatol Treat* 1: 199–201

2   Atherton DJ (1983) Breast feeding and atopic eczema. *Br Med J* 287: 775–776

3   Bamford JTM, Gibson RW, Renier CM (1985) Atopic eczema unresponsive to evening primrose oil (linoleic and γ-linolenic acids). *J Am Acad Dermatol* 13: 959–965

4   Barr SD, Brain S, Camp RDR, Cilliers J, Greaves MW, Mallet AI, Misch K (1984) Human allergic and irritant contact dermatitis: levels of arachidonic acid and its metabolites in involved skin. *Br J Dermatol* 111: 23–28

5   Bartel RL, Marcelo CL, Voorhees JJ (1987) Partial charcterization of phospholipase C activity in normal, psoriatic uninvolved and lesional epidermis. *J Invest Dermatol* 88: 447–451

6   Berth-Jones J, Graham-Brown RAC (1993) Placebo-controlled trial of essential fatty acid supplementation in atopic dermatitis. *Lancet* 341: 1557–1560

7   Biagi PL, Bordoni A, Hrelia S, Celadon M, Ricci GP, Cannella V, Patrizi A, Specchia F, Masi M (1994) The effect of gamma-linolenic acid on clinical status, red cell fatty acid composition and membrane microviscosity in infants with atopic dermatitis. *Drugs Exptl Clin Res* 20: 77–84

8   Biagi PL, Bordoni A, Masi M, Ricci G, Fanelli C, Patrizi A, Ceccolini E (1988) A long-term study on the use of evening primrose oil (efamol) in atopic children. *Drugs Exptl Clin Res* 14: 285–290

9   Bittiner SB, Tucker WF, Cartwright I, Bleehen SS (1988) A double-blind, randomized, placebo-controlled trial of fish oil in psoriasis. *Lancet* 331: 378–380

10 Bjorneboe A, Smith AK, Bjorneboe GE, Thune PO, Drevon CA (1988) Effect of dietary supplementation with n-3 fatty acids on clincal manifestation of psoriasis. *Br J Dermatol* 118: 77–83

11 Borelli S, Bresser H, Belsan I (1994) Externe Therapie mit γ-Linolensäure – Ergebnisse einer Doppelblindstudie. *Z Hautkr* 69: 523–524

12 Brain SD, Camp RDR, Dowd PM, Black AK, Wollard PM, Mallet AI, Greaves MW (1982) Psoriasis and leukotriene B$_4$. *Lancet* 321: 762–763

13 Buscino L, Ioppi M, Morse NL, Nisini R, Wright S (1993) Breast milk from mothers of children with newly developed atopic eczema has low levels of long chain polyunsaturated fatty acids. *J Allergy Clin Immunol* 91: 1134–1139

14 Camp RDR, Mallet AI, Wollard PM, Brain SD, Black AK, Greaves MW (1983) The identification of hydroxy fatty acids in psoriatic skin. *Prostaglandins* 26: 431–447

15 Cant A, Shay J, Horrobin DF (1991) The effect of maternal supplementation with linoleic and γ-linolenic acids on the fat composition and content of human milk: a placebo-controlled trial. *J Nutr Sci Vitaminol* 37: 573–579

16 Downing DT, Stewart ME, Wertz PW, Colton SW, Abraham W, Strauss JS (1987) Skin lipids: an update. *J Invest Dermatol* 88 (3 Suppl): 2s–6s

17 Downing DT, Stewart ME, Wertz PW, Strauss JS (1986) Essential fatty acids and acne. *J Am Acad Dermatol* 14: 221–225

18 Duell EA, Ellis CN, Voorhees JJ (1988) Determination of 5,12, and 15-lipoxygenase products in keratomed biopsies of normal and psoriatic skin. *J Invest Dermatol* 91: 446–450

19 Elias PM (1983) Epidermal lipids, barrier function and desquamation. *J Invest Dermatol* 80 (Suppl): 44s–49s

20 Elias PM (1981) Epidermal lipids, membranes and keratinization. *Int J Dermatol* 20: 1–19

21 Elias PM (1987) Plastic wrap revisited: the stratum corneum two-compartment model and its clinical implication. *Arch Dermatol* 123: 1405–1406

22 Elias PM, Brown BE (1978) The mammalian cutaneous permeability barrier: defective barrier function in essential fatty acid deficiency correlates with abnormal intercellular lipid deposition. *Lab Invest* 39: 574–583

23 Elias PM, Brown BE, Ziboh VA (1980) The permeability barrier in essential fatty acid deficiency: evidence for a direct role for linoleic acid in barrier function. *J Invest Dermatol* 74: 230–233

24 Elias PM, Goerke J, Friend DS (1977) Mammalian epidermal barrier layer lipids: composition and influence on structure. *J Invest Dermatol* 69: 535–546

25 Escobar SO, Achenbach R, Iannantuono R, Torem V (1992) Topical fish oil in psoriasis – a controlled and blind study. *Clin Exp Dermatol* 17: 159–162

26 Feingold KR, Brown BE, Lear SR, Moser Ahm Elias PM (1992) Effect of essential fatty acid deficiency on cutaneous sterol synthesis. *J Invest Dermatol* 87: 588–591

27 Fiocchi A, Sala M, Signoroni P, Baderali G, Agostoni C, Riva E (1994) The efficacy and safety of γ-linolenic acid in the treatment of intantile atopic dermatitis. *J Int Med Res* 22: 24–32

28 Fogh K, Herlin T, Kragballe K (1989) Eicosanoids in skin of patients with atopic dermatitis: prostaglandin E$_2$ and leukotriene B$_4$ are present in biologically active concentrations. *J Allergy Clin Immunol* 83: 450–455

29 Ford-Hutchinson AW, Bray MA, Smith MJH (1980) Lipoxygenase products and the polymorphonuclear leucocyte. *Agents Actions* 10: 548–550

30 Gibson RA, Kneebone GM (1981) Fatty acid composition of human colostrum and mature breast milk. *Am J Clin Nutr* 34: 252–257

31 Goldman DW, Pickett WC, Goetzel EJ (1983) Human neutrophil chemotactic and degranulating activities of leukotriene B$_5$ (LTB$_5$) derived from eicosapentaenoic acid. *Biochem Biophys Res Commun* 117: 282–288

32 Goldstein G, Scheid M, Hammerling U, Boyse EA, Schlesinger DH, Niall HD (1975) Isolation of a polypeptide that has lymphocyte differentiating properties and is probably represented universally in living cells. *Proc Natl Acad Sci USA* 72: 11–15

33 Goodwin JS, Ceuppen J (1983) Regulation of the immune response by prostaglandins. *J Clin Immunol* 3: 295–315

34 Goodwin JS, Bankhurst AD, Messner RP (1977): Suppression of human T-cell mitogenesis by prostaglandin. *J Exp Med* 146: 1719–1734

35 Grabbe J, Czarnetzki BM, Mardin M (1982) Chemotactic leukotrienes in psoriasis. *Lancet* 321: 1464

36 Grattan C, Burton JL, Manku M, Stewart C, Horrobin DF (1990) Essential-fatty-acid metabolites in plasma phospholipids in patients with ichthyosis vulgaris, acne vulgaris and psoriasis. *Clin Exp Dermatol* 15: 174–176

37 Grimminger F, Mayser P, Papavassilis C, Thomas M, Schlotzer E, Heuer KU, Fuhrer D, Hinsch KD, Walmrath D, Schill WB et al (1993) A double-blind, randomized, placebo-controlled trial of n-3 fatty acid based lipid infusion in acute, extended guttate psoriasis. Rapid improvement of clinical manifestations and changes in neutrophil leukotriene profile. *Clin Invest* 71: 634–643

38 Grubauer G, Feingold KR, Elias PM (1987) Relationship of epidermal lipogenesis to cutaneous barrier function. *J Lipid Res* 28: 746–752

39 Guenther L, Wexler D (1987) Evening primrose oil in the treatment of atopic dermatitis. *J Am Acad Dermatol* 17: 860

40 Gupta AK, Ellis CN, Goldfarb MT, Hamilton TA, Voorhees JJ (1990) The role of fish oil in psoriasis. A randomized, double-blind, placebo-controlled study to evaluate the effect of fish oil and topical corticosteroid therapy in psoriasis. *Int J Dermatol* 29: 591–555

41 Gupta AK, Ellis CN, Tellner DC, Anderson TF, Voorhees JJ (1989) Double-blind, placebo-controlled study to evaluate the efficacy of fish oil and low-dose UVB in the treatment of psoriasis. *Br J Dermatol* 120: 801–807

42 Hamarström S, Hamberg M, Samuelsson B, Duell EA, Stawiski M, Voorhees JJ (1975) Increased concentrations of nonsterified arachidonic acid, 12-L-hydroxy-5,8,10,14-eicosatetraenoic acid, prostaglandin E$_2$ and prostaglandin F$_2$ in epidermis of psoriasis. *Proc Natl Acad Sci USA* 72: 5130–5134

43 Hartop PJ, Prottey C (1976) Changes in transepidermal water loss and the composition of epidermal lecithin after applications of pure fatty acid triglycerides to the skin of essential fatty acid-deficient rats. *Br J Dermatol* 95: 255–264

44 Hederos C-A, Berg A (1996) Epogam evening primrose oil treatment in atopic dermatitis and asthma. *Arch Dis Child* 75: 494–497

45 Henneicke von Zepelin HH, Mrowietz U, Farber L, Bruck-Borchers K, Schober C, Huber J, Lutz G, Kohnen R, Christophers E, Welzel D (1993) Highly purified omega-3-polyunsaturated fatty acids for topical treatment of psoriasis. Results of a double-blind, placebo-controlled multicentre study. *Br J Dermatol* 129: 713–717

46 Holleran WM, Feingold KR, Mao-Qiang M (1991) Regulation of epidermal sphingolipid synthesis by permeability barrier function. *J Lipid Res* 32: 1151–1158

47 Hollmann J, Melnik B, Lee M-S, Hofmann U, Plewig G (1991) Stratum-corneum- und Nagellipide bei Patienten mit atopischer Dermatitis. *Hautarzt* 42: 302–306

48 Hollmann J, Michelsen S, Jansen T, Plewig G, Rippke F (1996) Einfluß einer hoch-

dosierten oralen Therapie mit Nachtkerzensamenöl auf die epidermalen Hautlipidfraktionen bei schwerem atopischem Ekzem. *Z Hautkr* 71: 115–120

49   Horrobin DF (1989) Essential fatty acids in clinical dermatology. *J Am Acad Dermatol* 20: 1045–1053

50   Imokawa G, Aba A, Jin K, Higaki Y, Kawashima M, Hidano A (1991) Decreased level of ceramides in stratum corneum of atopic dermatitis: an etiologic factor in atopic dry skin? *J Invest Dermatol* 86: 523–526

51   Imokawa G, Akasaki S, Minematsu Y, Kawai M (1989) Importance of intercellular lipids in water-retention properties of the stratum corneum: induction and recovery study of surfactant dry skin. *Arch Dermatol* 96: 523–526

52   Imokawa G, Kuno H, Kawai M (1991) Stratum corneum lipids serve as a bound-water modulator. *J Invest Dermatol* 96: 845–851

53   Janossy IM, Raguz JM, Rippke F, Schwanitz HJ (1995) Effekte einer 12,5%igen Nachtkerzensamenöl-Creme auf hautphysiologische Parameter bei atopischer Diathese. *Z Hautkr* 70: 498–502

54   Jin K, Higaki Y, Takagi Yl Higuchi K, Yada Y, Kawashima M, Imokawa G (1994) Analysis of beta-glucocerebrosidase and ceramidase activities in atopic and aged dry skin. *Acta Derm Venereol* 74: 337–340

55   Juto P (1980) Elevated serum immunoglobulin E in T cell-deficient infants fed cow's milk. *J Allergy Clin Immunol* 66: 402–407

56   Kettler AH, Baughn RE, Orengo IF, Black H, Wolf JE Jr (1988) The effect of dietary fish oil supplementation on psoriasis. Improvement in a patient with pustular psoriasis. *J Am Acad Dermatol* 18: 1267–1273

57   Kragballe K (1989) Dietary supplementation with a combination of n-3 and n-6 fatty acids (super gamma-oil marine) improves psoriasis. *Acta Derm Venereol* 69: 265–268

58   Kragballe K, Duell EA, Voorhees JJ (1986) Selective decrease of 15- hydroxyeicosate-traenoic acid (15-HETE) formation in uninvolved psoriatic dermis. *Arch Dermatol* 122: 877–880

59   Kragballe K, Fogh K (1989) A low-fat diet supplemented with dietary fish oil (Max-EPA) results in improvement of psoriasis and in formation of leukotriene B_5. *Acta Derm Venereol* 69: 23–28

60   Kragballe K, Voorhees JJ (1985) Arachidonic acid in psoriasis. Pathogenic role and pharmacological regulation. *Acta Derm Venereol Suppl* 120: 12–17

61   Kragballe K, Voorhees JJ, Goetzl EJ (1987) Inhibition by leukotriene B_5 of leukotriene B_4-induced activation of human keratinocytes and neutrophils. *J Invest Dermatol* 88: 555–558

62   Kromann N, Green A (1980) Epidemiological studies in the Upernavik District, Greenland. *Acta Med Scand* 208: 401–406

63   Kuester W, Petersen M, Christophers E, Goos M, Sterry W (1990) A family study of atopic dermatitis. Clinical and genetic characteristics of 188 patients and 2151 family members. *Arch Dermatol Res* 282: 98–102

64 Lans DM, Rocklin RE (1989) Dysregulation of arachidonic acid release and metabolism by atopic mononuclear cells. *Clin Exp Allergy* 19: 37–44

65 Liao S (1994) Androgen action: molecular mechanism and medical application. *J Formos Med Assoc* 93: 741–751

66 Macdonald KJS, Green C, Raffle EJ, Kenicer KJA (1985) Topical evening primrose seed oil and atopic eczema. *Scott Med J* 30: 267

67 Man MQM, Feingold KR, Thornfeldt CR, Elias PM (1996) Optimization of physiological lipid mixtures for barriere repair. *J Invest Dermatol* 106: 1096–1101

68 Manku MS, Horrobin DF, Morse N, Kyte V, Jenkins K, Wright S, Burton JL (1982) Reduced levels of prostaglandin precursors in the blood of atopic patients: defective delta-6-desaturase function as a biochemical basis for atopy. *Prostaglandins Leukot Med* 9: 615–628

69 Manku MS, Horrobin DF, Morse N, Wright S, Burton JL (1984) Essential fatty acids in the plasma phospholipids of patients with atopic eczema. *Br J Dermatol* 110: 643– 648

70 Maurice PD, Allen BR, Barkley AS, Cockbill SR, Stammers J, Bather PC (1987) The effect of dietary supplementation with fish oil in patients with psoriasis. *Br J Dermatol* 117: 599–606

71 Mead JF, Willia AL (1987) The essential fatty acids: their derivation and role. In: Willis AL (ed): *CRC handbook of eicosanoids: prostaglandins and related lipids*, vol 1, part A, CRC Press, Boca Raton, FL, 85–98

72 Melnik B (1989) Epidermal lipids and the biochemistry of keratinization. In: Traupe H (ed): *The ichthyoses*. Springer-Verlag, Berlin, 15–42

73 Melnik B, Braun-Falco O (1996) Bedeutung der Ölbäder für die adjuvante Basistherapie entzündlicher Dermatosen mit trockener, barrieregestörter Haut. *Hautarzt* 47: 665–672

74 Melnik B, Hollmann G, Plewig G (1988) Decreased stratum corneum ceramides in atopic individuals – a pathochemical factor in xerosis. *Br J Dermatol* 119: 547–549

75 Melnik B, Plewig G (1989) Ein neues Konzept zur Ätiopathogenese und Prävention der Atopie. *Hautarzt* 40: 685–692

76 Melnik B, Plewig G (1993) Externe Behandlung des atopischen Ekzems mit Gamma-linolensäure? *Hautarzt* 40: 604–605

77 Melnik B, Plewig G (1991) Immunregulatorische Bedeutung der E-Prostaglandine bei Atopie. *Hautarzt* 42: 211–214

78 Melnik B, Plewig G (1992) Modulation des Omega-6-Fettsäure- und Eikosanoidstoffwechsels bei atopischer Dermatitis. *Hautarzt* 43: 800–802

79 Morse PF, Horrobin DF, Manku MS, Stewart JCM, Allen R, Littlewood S, Wright S, Burton J, Gould DJ, Holt PH et al (1989) Meta-analysis of placebo-controlled studies of the efficacy of epogam in the treatment of atopic eczema. Relationship between plasma essential fatty acid changes and clinical response. *Br J Dermatol* 121: 75–90

80 Nissen HP, Wehrmann W, Kroll U, Kreysel HW (1988) Veränderungen im Plasma-Lipid-Muster bei Patienten mit Neurodermitis – Beeinflussung durch Applikation ungesättigter Fettsäuren. *Fat Sci Technol* 7: 268–271

81 Pillay DJ, Pope BL (1986) Requirement of prostaglandin E_1 (PGE_1) for the secretion of suppressor cell inducer factors by spleen cells of tumor-bearing mice. *Int J Immunopharmacol* 8: 221–226

82 Prottey C, Hartop PJ, Black JG, McCormack JI (1976) The repair of impaired epidermal barrier function in rats by the cutaneous application of linoleic acid. *Br J Dermatol* 94: 13–21

83 Ruzicka T (1985) Leukotriene und Monohydroxyfettsäuren: Kontroverse Rolle in der Pathogenese der Psoriasis. *Hautarzt* 36: 255–258

84 Ruzicka T (1984) Stoffwechsel der Arachidonsäure in der Haut und seine Bedeutung in der Pathophysiologie entzündlicher Dermatosen. *Hautarzt* 35: 337–343

85 Ruzicka T, Simmet T, Peskar BA, Ring J (1986) Skin levels of arachidonic acid-derived inflammatory mediators and histamine in atopic dermatitis and psoriasis. *J Invest Dermatol* 86: 105–108

86 Saarinen UM, Kajosaari M, Backman A, Siimes MA (1979) Prolonged breast-feeding as prophylaxis for atopic disease. *Lancet* 314: 163–166

87 Santoli D, Phillips PD, Colt TL, Zurier RB (1990) Suppression of interleukin 2- dependent human T cell growth by E-series prostaglandins (PGE) and their precursor fatty acids: evidence for a PGE-independent mechanism of inhibition by the fatty acids. *J Clin Invest* 85: 424–432

88 Santoli D, Zurier RB (1989) Prostaglandin E precursor fatty acids inhibit human IL-2 production by a PGE-independent mechanism. *J Immunol* 143: 1303–1309

89 Schaefer L, Kragballe K (1991) Supplementation with evening primrose oil in atopic dermatitis: effect on fatty acids in neutrophils and epidermis. *Lipids* 26: 557–560

90 Schalin-Karrila M, Mattila L, Jansen CT, Uotila P (1987) Evening primrose oil in the treatment of atopic eczema: effect on clinical status, plasma phospholipid fatty acids and circulating blood prostaglandins. *Br J Dermatol* 117: 11–19

91 Schneider I-M, Wohlrab W, Neubert R (1997) Fettsäuren und Epidermis. *Hautarzt* 48: 303–310

92 Scott MJ, Scott AM (1992) Effects of anabolic-androgenic steroids on the pilosebaceous unit. *Cutis* 50: 113–116

93 Sinclair HM (1958) Essential fatty acids and the skin. *Br Med Bull* 14: 258–262

94 Skogh M (1986) Atopic eczema unresponsive to evening primrose oil (linoleic and γ-linolenic acids). *J Am Acad Dermatol* 15: 114–115

95 Soyland E, Funk J, Rajka G, Sandberg M, Thune P, Rustad L, Helland S, Middlefart K, Odu S, Falk ES et al (1993) Effect of dietary supplementation with very-long-chain n-3 fatty acids in patients with psoriasis. *N Engl J Med* 328: 1812–1816

96 Stewart ME (1992) Sebaceous gland lipids. *Semin Dermatol* 11: 100–105

97 Stewart ME, Greenwood R, Cunliffe WJ, Strauss JS, Downing DT (1986) Effect of cytoproterone acetate-ethinyl estradiol treatment on the proportions of linoleic and sebaleic acids in various skin surface lipid classes. *Arch Dermatol Res* 278: 481–485

98 Stobo JD, Kennedy MD, Goldyne ME (1979) Prostaglandin E modulation of the mitogenic response of human T cells. *J Clin Invest* 64: 1188–1195

99 Thestrup-Pederson K, Halkier-Sorensen L (1992) Skin physiological parameters in patients with atopic dermatitis before and after the use of emollients with and without essential fatty acids (poster). 18th World Congress of Dermatology, New York City, 12–18 June

100 Veale DJ, Torley HI, Richards IM, O'Dowd A, Fitzsimons C, Belch JJ, Sturrock RD (1994) A double-blind placebo controlled trial of Efamol Marine on skin and joint symptoms of psoriatic arthritis. *Br J Rheumatol* 33: 854–958

101 Walker C, Kristensen P, Bettens F, DeWeck AL (1983) Lymphokine regulation of activated (G_1) lymphocytes. I. Prostaglandine E_2-induced inhibition of interleukin 2 production. *J Immunol* 130: 1770–1773

102 Webb DR, Wieder KJ, Rogers TJ, Healy CT, Nowowiejski-Wieder I (1985) Chemical identification of prostaglandin-induced T suppressor (PITS). *Lymphokine Res* 4: 139–149

103 Werner Y, Lindberg M, Forslind B (1982) The water-binding capacity of stratum corneum in dry non-eczematous skin of atopic eczema. *Acta Derm Venereol* 62: 334–337

104 Werner Y, Lindberg M (1985) Transepidermal water loss in dry and clinically normal skin in patients with atopic dermatitis. *Acta Derm Venereol* 65: 102–105

105 Wertz PW, Cho ES, Downing DT (1983) Effect of essential fatty acid deficiency on the epidermal sphingolipids of the rat. *Biochem Biophys Acta* 753: 350–355

106 Wertz PW, Downing DT (1982) Glycolipids in mammalian epidermis: structure and function in the water barrier. *Science* 217: 1261–1262

107 Wertz PW, Miethke MC, Long SA, Strauss JS, Downing DT (1985) The composition of the ceramides from human stratum corneum and from comedones. *J Invest Dermatol* 84: 410–412

108 Wertz PW, Swartzendruber DC, Abraham W, Madison KC, Downing DT (1987) Essential fatty acids and epidermal integrity. *Arch Dermatol* 123: 1381–1384

109 Wong E, Camp R, Greaves M (1984). Topical application of leukotriene B_4 in psoriatic and normal subjects. *J Invest Dermatol* 82: 414

110 Wright S, Bolton C (1989) Breast milk fatty acids in mothers of children with atopic eczema. *Br J Nutr* 62; 693–697

111 Yamamoto G, Serizawa S, Ito M, Sato Y (1991) Stratum corneum lipid abnormalities in atopic dermatitis. *Arch Dermatol* Res 283: 219–223

112 Yamamoto A, Takenouchi K, Ito M (1995) Impaired water barrier function in acne vulgaris. *Arch Dermatol* Res 287: 214–218.

113 Ziboh VA, Cohen KA, Ellis CN, Miller C, Hamilton TA, Kragballe K, Hydrick CR, Voorhees JJ (1986) Effects of dietary supplementation of fish oil on neutrophil and epidermal fatty acids. Modulation of clinical course of psoriatic subjects. *Arch Dermatol* 122: 1277–1282

114 Zurier RB (1993) Fatty acids, inflammation and immune response. *Prostaglandins Leukot Essent Fatty Acids* 48: 57–62

Inhibitors of eicosanoid biosynthesis in skin inflammation

Hans F. Merk

Hautklinik der Medizinischen Fakultät der RWTH Aachen, Pauwelsstr. 30, D-52074 Aachen, Germany

Introduction

Inflammation is a characteristic of many dermatological disorders. The skin contains multiple types of epidermal and dermal cells, and cells mediating inflammatory reactions are present in the skin during inflammation and are able to release and to metabolize arachidonic acid (AA) [7]. Several derivatives of AA have been shown to be major mediators of skin inflammation [23]. Substances such as glucocorticosteroids or nonsteroidal antiinflammatory drugs (NSAIDS), which interfere with the release or metabolism of AA, are used for the downregulation of inflammatory reactions. In particular, it has been shown that keratinocytes produce derivatives of the cyclooxygenase, 5-, 12- and 15-lipoxygenase pathway [31]. Cyclooxygenase (COX) catalyses the synthesis of cyclic endoperoxides from AA to form prostaglandins (PGs). COX is a bifunctional enzyme that catalyzes the bisoxygenation of AA to form PGG_2 and a peroxidase activity, which catalyze the reduction of PGG_2 to PGH_2. This enzyme was considered to be unique. However, in recent years it has been shown that COX activity can be induced in a variety of cells after exposure to endotoxins, proinflammatory cytokines, growth factors, hormones or tumor promoters [17]. This induction, which can be inhibited by glucocorticoids, led to the discovery that there is a constitutive COX activity – COX1 – and an inducible COX activity – COX2. These enzymes are encoded by different genes which are distinctively regulated at both the transcriptional and posttranscriptional level [32]. The promoter of the *COX1* gene is consistent with a housekeeping gene and poorly inducible, whereas *COX2* gene possesses promoters with many transcriptional factor consensus sequences. The 3'-untranslated region of COX2 has several copies of Shaw-Kamen's sequence, which is found in many immediate-early genes and confers enhanced messenger RNA (mRNA) degradation [30, 22, 33]. These characteristics underline that COX2 is inducible and well regulated. Studies with COX1 and COX2 knockout mice showed a reduced life span for the COX2 knockouts, which had a normal inflammatory response to AA and phorbolester stimulation, whereas COX1 knockouts had a decreased inflammatory response to AA, but not to phorbol ester stimulation [20, 29, 10]. COX1 as well as COX2 have been shown to be present in murine and human keratinocytes, and a multifactor regulator system of COX2 in murine keratinocytes has been characterised by compounds such as phorbol esters, anthralin, and benzoylperoxide or ultraviolet (UV)-light [25, 16, 21, 19].

Other mediators of inflammation in the skin include the platelet-activating factor (PAF), an ether phospholipid and, like AA, produced by the action of phospholipase A_2 and acetyl transferase. Sources of the different mediators are poorly understood. Skin cells other than keratinocytes such as Langerhans' cells produce these metabolites as well, although they differ from keratinocytes in that they form especially PGD_2 and 12-HETE [19].

AA is released by hydrolytic cleavage from its ester linkage to the 2-acyl position of cell membrane lipids. Oleate or linoleate can also be linked to this position, which has been used in special diets in an effort to prevent excessive AA release. Several enzymes are involved in this hydrolysis. The first involves activation of diacylglycerol (DAG) as second messenger by separating phosphatidylinositol (IP_3) and DAG from phosphatidylinositol 4,5-bisphosphate (PIP) after binding of an agonist to the G protein receptor or after expression of the Harvey-ras-oncogene [1, 3]. This results in the activation of phospholipase C and a highly specific phospholipase A_2, which liberates AA in the 2-position of the phosphoinositide. A second pathway involves a less specific phospholipase A_2 which cleaves AA and requires a high amount of Ca^{2+} [4]. Various phospholipids such as phosphatidylcholine or phosphatidylethanolamine are involved in this reaction [4]. The importance of the first pathway in the liberation of AA remains unclear, but it is known to exist in human skin [7]. The liberation of arachidonic acid is also controlled by lipocortins, which among other antiinflammatory effects inhibit phospholipase A_2 activity [13].

NSAIDs

Three main NSAID groups can be distinguished by their pharmacological properties: the acidic antiinflammatory NSAIDs, the pyrazolones and acetaminophen (Tab. 1). The aspirin-like drugs are characterized by their ability to inhibit COX or prostaglandin synthetase (Fig. 1). They are most useful as analgesics in severe inflammatory reactions such as rheumatoid arthritis or UV-erythema [25]. In dermatology, most experience exists with aspirin, indomethacine, meclofenamate and bufexamac, although they have all proven to be of little therapeutic benefit in dermatology [15]. All conventional NSAIDs including aspirin are nonselective inhibitors of COX1 and COX2. Up to now there is no experience with regard to skin pharmacology with selective COX2 inhibitors such as sulfonamides, e.g. L-745,337, or tricyclic methyl sulfone derivatives, e.g. SC58125. Salicylate itself is a very weak competitive inhibitor of COX activity and should be considered as a non-

Table 1 - Major classes of compounds

Acid antiinflammatory NSAID	Salicylates
	Mefenamic acid
	Ibuprofen
	Indomethacin
	Bufexamac
Pyrazolones	
Acetaminophen	

anti-COX drug, although it is derived from aspirin and may be responsible for its chronic antiinflammatory activity [17]. Interestingly, recent observations indicated that salicylate suppresses COX2 and COX1 expression in cultured cells and reduces the steady-state level of COX mRNA in cultured umbilical vein endothelial cells [37, 38]. Of special interest is acetaminophen, which primarily influences COX-activity in the central nervous system (CNS) and possesses low activity on COX1 and COX2 in other cells [17].

Acetylsalicylic acid

Since PGE has been shown to decrease the threshold of histamine-evoked cutaneous itching, acetylsalicylic acid (ASA) has been used as an antipruritic agent. There are some case reports of its effectiveness in the pruritus of atopic dermatitis [15]. Its effectiveness was also proven in a double-blind crossover study in patients with pruritus associated with polycythemia rubra vera. Daily administration of 1.5 g to 17 patients resulted in improvement in 15 of 17 patients [12]. However, this observation was not confirmed in a study in patients with aquagenic pruritus associated with polycythemia [15]. Oral ASA reduces UV-light-induced erythema [25]. Furthermore it has been shown to be helpful in systemic mastocytosis when given together with H_1 and H_2-antihistamines [27].

Indomethacin

Indomethacin is a methylated indoleacetic acid derivative. Due to its wide use in several dermatopharmacologic studies, it became a prototypical NSAID in addition to aspirin. It inhibits both COX and reduction of hydroperoxyeicosatetraenoic acid (HPETE) to hydroxyeicosatetraenoic (HETE) acid. This results in an accumulation of HPETE, which can also inhibit prostaglandin synthetase [13]. Thus indomethacin

93

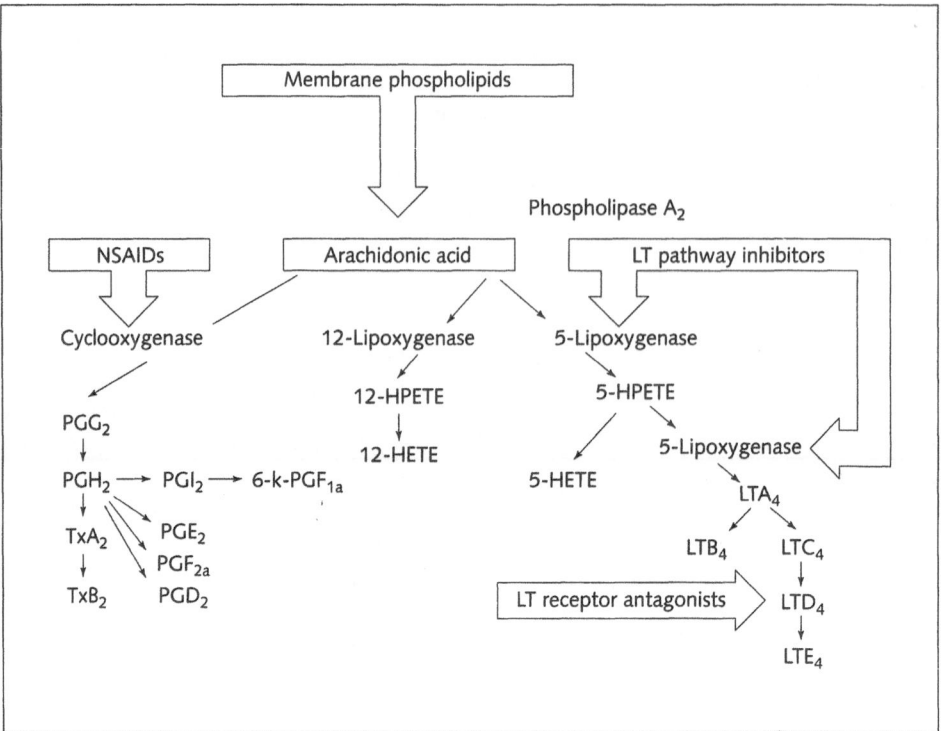

Figure 1
Arachidonic acid is released by different pathways. After binding of an agonist to the G pro-
tein or a product of the ras-oncogene DAG is formed from PIP2. DAG can be further metab-
olized to monoacylglycerol and arachidonic acid, which is also formed by phospholipase C-
dependent activation from phosphatidylinositol. An alternative pathway involves phospho-
lipase A, which can be inhibited by glucocorticoids via lipocortins. AA is further
metabolized to prostaglandins and leukotrienes. Only the pathway to prostaglandins, which
is mediated by COX1 and COX2, is inhibited by the available NSAID. The potency of dif-
ferent NSAIDs depends on their inhibition of prostaglandin synthetase. Aspirin inhibits this
pathway irreversibly, whereas the other drugs inhibit reversibly. New drugs which may play
a role in the treatment of asthma inhibit the leukotriene (LT) pathway or are LT receptor
antagonists [36].

inhibits COX in at least two ways. Indomethacin has been studied in different types
of cutaneous inflammation, including UV erythema, dithranol erythema and psoriasis, and epidermal and dermal wounds [15]. Its effectiveness varies in these different inflammatory models. Systemically and topically applied indomethacin inhibits
or diminishes reduced erythema due to UVB (290 to 320 nm) or UVC (253 nm)

radiation; this finding was paralleled by decreased formation of PGE_2 and $PGF_{2\alpha}$, and similar results were obtained with other NSAIDs such as flubiprofen [15]. The topical application of indomethacin (2.5%) proved to be more effective than a topical corticosteroid (fluocinolone acetonide 0.05%); a similar finding was obtained with ibuprofen (5%) [34]. However, 24 h after irradiation erythema was not inhibited by indomethacin, which may be related to the proinflammatory agent 12-HETE, another derivative of arachidonic acid whose formation is not prevented by indomethacin [14].

The results concerning dithranol erythema are conflicting. Some investigators found an increase after using indomethacin, others found a decrease. In psoriasis, the topical application of indomethacin cream (1%) as well as meclofenamate cream (1%) worsened psoriasis in 14 out of 20 and 7 out of 10 patients, respectively [11]. The authors proposed that the nonsteroidal antiinflammatory drugs contributed to the maintenance of the patients' psoriasis by altering the AA transformation cascade through leukotriene production. Interestingly the antipsoriatic agent dithranol reduces the amount of AA in psoriatic lesions without influencing LTB_4 and 12-HETE [18]. Wound healing is not influenced by indomethacin or other NSAIDs such as ibuprofen or meclofenamate sodium; however, erythema, heat and edema after mechanical injury of the skin is reduced [2].

In conclusion, these data show that in assessing NSAID efficacy, it is necessary to specify the cutaneous inflammation reaction being studied. This is supported by the observation that different AA derivatives and other proinflammatory mediators including PAF and interleukin-1 (IL-1) or active oxygen species can participate in different inflammation reactions of the skin [14]. There are isolated reports of the effectiveness of indomethacin (25–50 mg up to 100 mg/day) in urticarial vasculitis and in delayed pressure urticaria; however, we have been unimpressed by the use of indomethacin or other NSAIDs in these diseases. Erythema nodosum is another disease for which indomethacin is recommended. It has been suggested that prostaglandin formation is especially important for the intense inflammatory reaction in subcutaneous fat. In patients suffering from high fever during erythema nodosum, pyrazolone derivatives may be helpful, but care must be taken to be sure that they are not the cause of the disease.

Bufexamac

The NSAID (p-butoxyphenyl)acetohydroxamic acid is widely recommended as an ointment preparation in several European countries for inflammatory reactions such as allergic and irritant dermatitis, perianal pruritus, UV erythema, atopic dermatitis and even psoriasis. However, the available clinical studies are far from convincing since they were not performed in a double-blind controlled fashion and do not distinguish several different kinds of inflammatory reactions [15, 27].

Taken together, the available NSAIDs play no major role in dermatological therapy. The main reason appears to be that they only inhibit the COX pathway. Therefore, new drugs which inhibit other lipoxygenase pathways or special receptors of derivatives of AA may prove more useful in the future.

Side effect

The main cutaneous side effects of NSAIDs are pseudo-allergic reactions, especially to acetylsalicylic acid, allergic reactions to pyrazolone and photosensitivity reactions (Tab. 2). Also, a pemphigoid-like bullous eruption related to ibuprofen has been described [21].

ASA is estimated to be the cause of 5 to 10% of acute or chronic urticaria and worsens this disease in 22 to 40% of affected individuals [26]. There are several reasons why this kind of reaction is not allergic but pseudo-allergic:

1) There are cross-reactions in patients with this disease to other NSAIDs although they have quite another chemical structure.
2) There are no immunoglobulin E (IgE) antibodies found to ASA or to other NSAIDs which might explain these reactions.
3) There are no reactions to aspirin in the prick or intracutaneous (i.c.) test in these patients [26].

There is some evidence that these reactions are mediated by alteration of the metabolism of AA by NSAIDs. There are some observations, that ASA can inhibit the Cl-esterase inactivator, which might increase the production of C3a and C5a, known

Table 2 - Cutaneous side effects to NSAIDs

Salicylates	Acute urticaria
	Chronic urticaria
	Purpura (aspirin)
	Exanthem
Acid antiinflammatory NSAID	Exanthem
	Erythema multiforme-like eruption
	TEN
	Anaphylactoid reaction
	Photosensitivity
	Fixed drug eruption
	Pruritus

to induce histamine release from basophils and mast cells [35]. Decreased glutathione peroxidase was found in lymphocytes of patients with a pseudo-allergic reaction to aspirin [24]. This enzyme plays a role in the metabolism of potential toxic compounds as well as in the metabolism of AA derivatives.

The development of leukotriene-receptor antagonists for the treatment of asthma – including ASA-induced asthma – as well as further investigations of the metabolism of AA under the influence of ASA and other NSAIDs revealed following results [9]:

1) Those patients have an increased excretion of leukotriene E_4 in the urine under normal conditions without stimulation.
2) They produce leukotrienes after oral, nasal or bronchial challenge with ASA or lysine-ASA.
3) *In vitro* their basophils secrete after stimulation with C5a and/or lysine-ASA an increased amount of leukotriene derivatives.
4) The pretreatment with specific leukotriene-receptor antagonists results in prevention of asthma by ASA.

These observations suggested that, under normal conditions, the formation of PGE_2 beside leukotrienes is a protection that is lost after the ingestion of NSAIDs resulting in these side effects [9].

Cutaneous side effects of pyrazolone derivatives include pseudo-allergic or allergic reactions [26]. In severe cases of anaphylactic reactions to these compounds or in bullous reactions such as toxic epidermal necrolysis or fixed drug reactions, immunologic mechanisms are involved [5]. In one patient who also developed severe anaphylactic shock after ingesting propyphenazone, we performed a lymphocyte transformation test with the drug itself and with a drug-modified microsomal preparation [28]. In this case we incubated propyphenazone with cytochrome P-450 containing murine microsomes prior to performing the lymphocyte transformation test. After this preincubation, we observed increased proliferation and blast transformation of the lymphocytes, indicating that this patient was not sensitized by the drug itself, but rather by a metabolite. Further cases with similar results are summarized in Figure 2.

Acetaminophen, which is often recommended for patients with pseudo-allergic reactions to NSAIDs and as an alternative to ASA for the relief of minor itching, pain and febrile disorders, in rare cases induces fulminant hepatic necrosis [6]. This reaction is a toxic one and occurs after the ingestion of at least 15 g, which means a dose of about 50 tablets. However, some patients also react after only 3 or 4 g [6]. The cause of this reaction is the metabolism of the drug. The active metabolite is most likely N-acetylimidoquinone, which binds to glutathione or to membranes, thereby inducing necrosis. This metabolite is formed by a cytochrome P-450 isoenzyme which is inducible by ethanol (P-450 2E). Therefore, alcoholics are at special

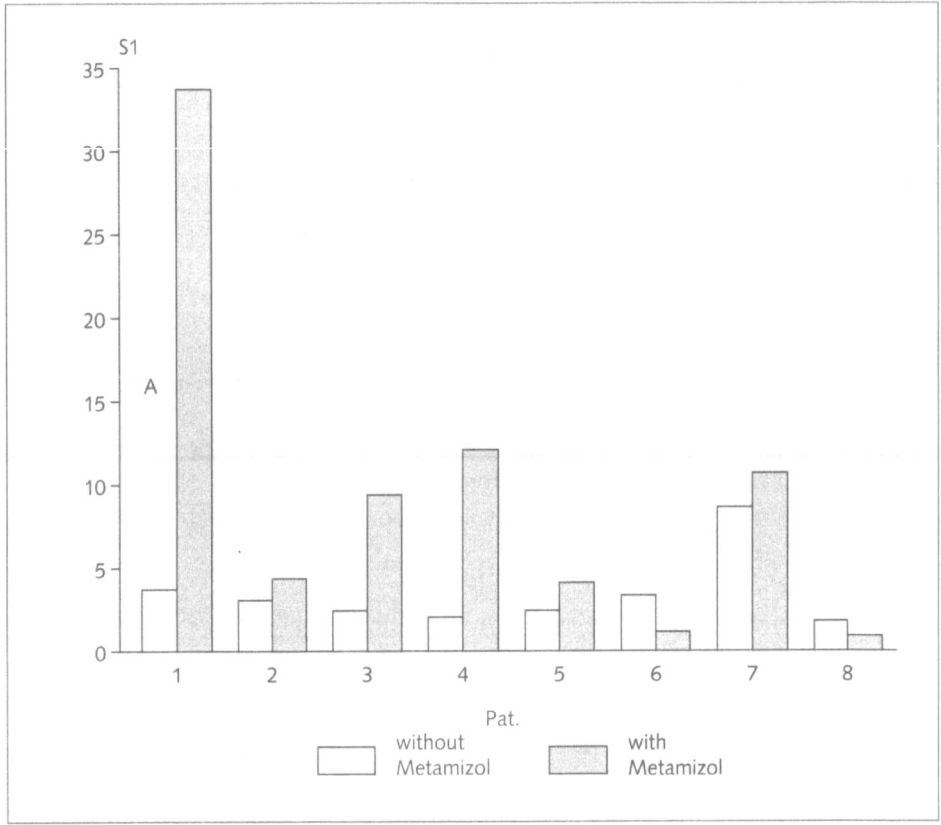

Figure 2
Different amounts of propyphenazone (15 to 150 µg/ml) were incubated with lymphocytes recently described with and without murine liver microsomes (M) known to metabolize this NSAID. The stimulation is given as stimulation index (S1). There was a positive reaction only after the addition of microsomes by the lymphocytes of the patient. Interestingly, there was a positive reaction to propyphenazone itself in the skin test as well [37].

risk to develop this type of hepatotoxicity. Also, patients who are fasting or who suffer from diabetes mellitus have increased cytochrome P-450 2E activity. Finally, patients receiving drugs which decrease the activity of glutathione transferase, for example azathioprine, are at high risk [6]. N-acetylcysteine may be helpful in these cases since it binds the highly reactive metabolite of acetaminophen similar to glutathione, and thereby functions as an antidote. However, it is helpful only for a short period after ingestion of the drug.

References

1   Agroff BW (1986) Inositol triphosphate and related metabolism. *Fed Proc* 45: 2627

2   Alvarez OM, Levendorf KD, Smerbeck RV, Mertz PM, Eagelstein, WH (1984) Effect of topically applied steroidal and nonsteroidal antiinflammatory agents on skin repair and regeneration. *Fed Proc* 43, 2793

3   Bell RM (1986) Protein kinase C activation by diacylglycerol second messengers. *Cell* 45: 631

4   Berridge MJ (1984) Inositol triphosphate and diacylglycerol as second messengers *Biochem J* 220: 345

5   Bigby M, Stern R (1985) Cutaneous reactions to nonsteroidal antiinflammatory drugs. *J Am Acad Dermatol* 12: 866

6   Black M (1980) Acetaminophen hepatotoxicity. *Gastroenterology* 78, 382

7   Camp RDR, Greaves MW (1987) Inflammatory mediators in the skin. *Br Med Bull* 43: 401

8   Chung KF, Holgate ST (1997): Leukotrienes: why are they important mediators in asthma? *Eur Resp Rev* 7: 259–263

9   Dahlen B, Dahlen SE (1995) Intolerance reactions to NSAIDs. In: A Basomba, J Sastre (eds): *Proceedings of the XVI European Congress of Allergology and Clinical Immunology ECACI 1995*. Monduzzi Editore, Madrid, 821–828

10 Dinchuk JE, Car BD, Focht RJ et al (1995) Renal abnormalities and an altered inflammatory response in mice lacking cyclooxygenase II. *Nature* 378: 406–409

11 Ellis CN, Fallon JD, Kang S, Vanderveen EE and Voorhees JJ (1986) Topical application of nonsteroidal antiinflammatory drugs prevents vehicle-induced improvement of psoriasis. *J Am Acad Dermatol* 14, 39

12 Fjellner B, Hagermark O (1979) Pruritus in polycythaemia vera: treatment with aspirin and possibility of platelet involvement. *Acta Dermatol Venereol* 59, 505

13 Flower RJ, Moncada S, Vane RV (1985) Analgesic-antipyretics and antiinflammatory agents: Drugs employed in the treatment of gout. In: *Goodman & Gilman's The Pharmacological basis of therapeutics*, 7th edn. Macmillan Publishing Company, New York, 674–715

14 Greaves MW, Camp RDR (1988) Prostaglandins, leucotrienes, phospholipase, platelet activating factor, and cytokines: An integrated approach to inflammation of human skin. *Arch Dermatol Res* 280, 33–41

15 Greaves MW (1987) Pharmacology and significance of nonsteroidal antiinflammatory drugs in the treatment of skin diseases. *J Am Acad Dermatol* 16, 751

16 Gresham A, Masferrer J, Chen X et al (1996) Increased synthesis of high molecular weight cPLA2 mediates early UV-induced PGE_2 in human skin. *Am J Physiol* 270: C1037–1050

17 Jouzeau JY, Terlain B, Abid A, Nédédec E, Netter P (1997) Cyclo-oxygenase isoenzymes. *Drugs* 53: 563–582

18 Juhlin L (1981) Factors influencing anthralin erythema. *Br J Dermatol* 105 (Suppl 20): 87

19 Katiyar SK, Mukhtar H (1997) Inhibition of phorbol ester tumor promoter 12-O-tetradecanoylphorbol-13-acetate caused inflammatory responses in SENCAR mouse skin by black tea polyphenols. *Carcinogenesis* 18: 1911–1916

20 Langenbach R, Morham SG, Tiano HF et al (1995) Prostaglandin synthase I gene disruption in mice reduces arachidonic acid-induced inflammation and indomethacin-induced gastric ulceration. *Cell* 83: 483–492

21 Leong J, Hughes-Fulford M, Rakhlin N et al (1996) COX in human and mouse skin and cultured human keratinocytes association of COX-2 expression with human keratinocyte differentiation. *Exp Cell Res* 224: 79–87

22 Lyons-Giordano B, Pratta MA, Galbraith W et al (1993) Interleukin-1 differentially modulates chondrocyte expression of cyclooxygenase-2 and phospholipase A_2. *Exp Cell Res* 206: 58–62

23 Majerus PW, Connolly TM, Deckmyn H, Ross TS, Bross TE, Ishii H, Bansal VS, Wilson DB (1986) The metabolism of phosphoinstide-derived messenger molecules. *Science* 234: 1519

24 Malgrem R, Unge G, Zetterström O, Theorell H, de Wahl K (1986) Lowered glutathioneperoxidase activity in asthmatic patients with food and aspirin intolerance. *Allergy* 41: 43

25 Maldve RE, Fischer SM (1996) Multifactor regulation of prostaglandin H synthase-2 in murine keratinocytes. *Mol Carcinogenesis* 17: 207–216

26 Merk H, G Goerz (1983): Analgetika-Intoleranz. *Z Hautkr* 58:535-542

27 Merk HF (1991) Glucocorticosteroids and NSAID – Mechanisms of action and their role in the treatment of skin diseases. In: T Ruzicka (ed): *Eicosanoids and the skin*. CRC Press, Boca Raton, FL, S171–184

28 Merk HF, Niederau D, Hertl M, Jugert F (1991) Drug metabolism and drug allergy. In: J Ring, B Przybilla (eds): *New trends in allergy III*. Springer-Verlag, Heidelberg, 269–280

29 Morham SG, Langenbach R, Loftin CD et al (1995) Prostaglandin synthase 2 gene disruption causes severe renal pathology in the mouse. *Cell* 83: 473–482

30 Ristimäki A, Garfinkel S, Wessendorf J et al (1994) Induction of cyclooxygenase-2 by interleukin-1. *J Biol Chem* 269: 11769–11775

31 Ruzicka T, Printz MP (1984) Arachidonic acid metabolism in skin: a review. *Rev Physiol Biochem Pharmacol* 100: 122

32 Smith WL, deWitt DL (1996) Prostaglandine endoperoxide H synthases-1 and -2. *Adv Immunol* 62: 167–215

33 Srivastava KS, Tetsuka T, Daphna-Iken D et al (1994) IL1β stabilizes COXII mRNA in renal mesangial cells: role of 3*-untranslated region. *Am J Physiol* 267: F504–508

34 Tan P, Flowers FP, Araujo OE, Doering P (1986) Effect on topically applied flubiprofen on ultraviolet-induced erythema. *Drug Int Clin Pharm* 20: 496

35   Voigtländer V, Hänsch GM, Rother U (1982) Intolerance to acetylsalicylic acid (aspirin) and tartrazine. *Arch Dermatol Res* 274: 359

36   Wenzel SE (1998) Investigational agents for the treatment of asthma. *J Allergy Clin Immunol* 101: S421–S423

37   Wu KK (1998) Biochemical pharmacology of nonsteroidal antiinflammatory drugs. *Biochem Pharmacol* 55: 543–547

38   Wu KK, Sanduja R, Tsai AL, Ferhanoglu B, Loose-Mitchell DD (1991) Aspirin inhibits interleukin-1 induced prostaglandin H synthase expression in cultured endothelial cells. *Proc Natl Acad Sci USA* 88: 2384–2387

Keratinocytes as a cellular source of inflammatory eicosanoids

Luis Vila, Rosa Antón and Mercedes Camacho

Laboratory of Inflammation Mediators, Institute of Research of the Santa Creu and Sant Pau Hospital, S. Antonio Mª Claret 167, E-08025 Barcelona, Spain

Introduction

The term 'eicosanoid' was introduced by Corey et al. [1] and comprises a large and complex family of compounds derived from 20-carbon polyunsaturated fatty acids, among which arachidonic acid is the most biologically relevant. This group of substances includes prostaglandins, thromboxanes, leukotrienes, hydroperoxy- and hydroxyeicosatetraenoic acids (HPETEs and HETEs), hepoxilins, lipoxins, trioxilins, nonleukotriene dihydroxyeicosatetraenoic acids (DiHETEs) and isoprostanes.

The biosynthesis of most eicosanoids is initiated with tightly controlled peroxidation of polyunsaturated fatty acids released from membrane lipids by the action of phospholipases. The abstraction of a hydrogen atom from a double allylic methylene group takes place in the molecule of substrate, yielding a carbon centred radical, which tends to be stabilized by a molecular rearrangement to form conjugate dienes. A further reaction of the lipid radical (R·) with molecular oxygen gives peroxyl radicals (ROO·). These are enzyme-bound intermediates in the formation of the pivotal compounds in the eicosanoid biosynthetic pathways. The formation of the peroxyl radicals may be catalysed by several lipoxygenases and prostaglandin H synthases (PGHS, also called cyclooxygenases). These enzymes recognise C7, C10 and C13 centered 1-*cis*, 4-*cis*-pentadiene structures in the molecule of arachidonic acid, to stereoselectively insert one molecule of oxygen at C5, C12 and C15 or C11, respectively. The insertion of an oxygen molecule at C5, C12 and C15 is catalysed by 5-, 12- and 15-lipoxygenase, respectively. The PGHS-dependent catalysis comprises two steps: during the first step, an oxygen molecule is inserted at C11, and in the second step, another oxygen is inserted at C15. The peroxyl radical can either be transformed into a hydroperoxide (ROOH) by picking up a hydrogen atom, or it can be added to a double bond to form cyclic peroxides termed prostaglandin endoperoxides. The former reaction is catalysed by lipoxygenases and the latter by PGHS (reviewed in [2]).

Hydroperoxides and prostaglandin endoperoxides undergo further reactions, including reduction by peroxidases to yield hydroxides (HETEs), dehydration to

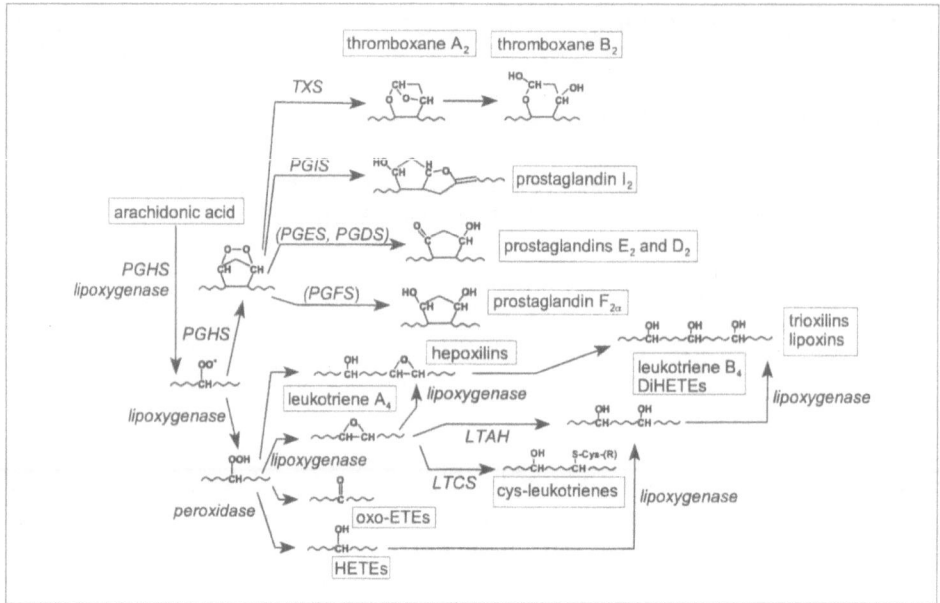

Figure 1

Simplified diagram of the different eicosanoid biosynthetic pathways. Enzymes are in italics, and typical compounds are framed. Brackets indicate that the reaction can be spontaneous or enzymatically catalysed. LTAH, leukotriene A$_4$ hydrolase; LTCS, leukotriene C$_4$ synthase; PGDS, prostaglandin D synthase; PGES, prostaglandin E synthase; PGFS, prostaglandin F synthase; PGHS, prostaglandin H synthase; PGIS, prostaglandin I synthase; TXS, thromboxane synthase.

yield carbonylic or epoxy derivatives such as oxo-ETEs or leukotriene A$_4$, respectively, and molecular rearrangements towards epoxy-hydroxy derivatives such as hepoxilins (reviewed in [3]). Endoperoxides may undergo spontaneous or enzymatic cleavage of O–O bonds to be transformed into prostaglandins and thromboxanes (reviewed in [4]).

Epoxy compounds can either undergo spontaneous or enzyme-catalysed water nucleophilic attack, or conjugate with glutathione to yield dihydroxy compounds such as leukotriene B$_4$ or peptidohydroxy derivatives such as leukotriene C$_4$, respectively. Hydroxy compounds can undergo the action of a lipoxygenase to yield DiHETEs. A lipoxygenase can also act on epoxy derivatives to yield epoxy-hydroxy derivatives. Dihydroxy and epoxy-hydroxy derivatives undergo further oxidations to yield trihydroxy derivatives such as lipoxins (reviewed in [5]) and trioxilins (reviewed in [6]). All these transformations are outlined in Figure 1.

Despite the fact that all mammalian cell types are able to produce eicosanoids, different cell types produce different eicosanoids from the qualitative and quantitative point of view. Such "specialized" ability to synthesise eicosanoids is regulated in some cases by restricted expression of the "secondary" enzymes. This could be the case of the prostanoids (prostaglandins and thromboxanes) formed through the PGHS pathway. As PGHS expression is so widespread, prostanoid biosynthesis by a particular cell type is determined by the expression of enzymes such as thromboxane synthase or prostaglandin I synthase. The opposite occurs in the biosynthesis of leukotrienes, which is limited by the scarce expression of the "primary" enzyme 5-lipoxygenase. Another way to control the biosynthesis of a particular eicosanoid is the concurrence of several enzymes supplied by different cell types in a cooperative way, as occurs with lipoxins [5]. The biosynthesis of eicosanoids by keratinocytes will be discussed below, with special attention to the main pathways through which they are formed: 5-lipoxygenase, PGHS and 12-lipoxygenase pathways.

Primary oxygen insertion in the arachidonic acid molecule at C5: the 5-lipoxygenase pathway

Enzymes involved in the 5-lipoxygenase pathway

The action of 5-lipoxygenase on arachidonic acid provides leukotriene A_4, which is a pivotal intermediate in the synthesis of leukotrienes. 5-Lipoxygenase is a bifunctional enzyme that requires Ca^{++} and adenosine triphosphate (ATP) for its activity. It generates 5-HPETE from arachidonic acid as an intermediate and leukotriene A_4 in a subsequent reaction. 5-HPETE may, alternatively, be reduced to 5-HETE (reviewed in [7]). In resting granulocytic cells, 5-lipoxygenase is located in the cytosol with a substantial proportion being concentrated in the nucleus. The transient increase of cytosolic Ca^{++} results in the translocation of 5-lipoxygenase from the cytosol to the cell membrane compartments to bind the 5-lipoxygenase-activating protein (FLAP), which favours the transfer of substrate to the enzyme. In activated cells both FLAP and 5-lipoxygenase are located in the nuclear envelope. Human and rat 5-lipoxygenase and FLAP have been cloned. 5-Lipoxygenase is inactivated after transforming the substrate when it is membrane-associated (reviewed in [7–9]). It has been observed that the 5-lipoxygenase activity is preserved, even in the membrane compartment, when substrate transformation is avoided by the presence of a reversible 5-lipoxygenase inhibitor [10].

Leukotriene A_4 is converted into leukotriene B_4 by leukotriene A_4 hydrolase, a cytosolic enzyme that also exerts aminopeptidase activity and contains zinc at its catalytic centre. The enzyme appears to be ubiquitously distributed and requires no additional cofactors for optimal activity (reviewed in [8, 11]).

Leukotriene C_4 synthase catalyses the conjugation of glutathione to C6 of leukotriene A_4 to yield leukotriene C_4. The enzyme is a unique member of the glutathione-S-transferase family. It cannot utilize xenobiotics as substrates (reviewed in [8, 12]). Leukotriene C_4 and its derivatives leukotrienes D_4 and E_4 are termed "peptidoleukotrienes" or "cysteinyl leukotrienes" because they contain a cysteine residue as a common constituent.

Keratinocyte 5-lipoxygenase pathway

Biological effects of leukotrienes, such as neutrophil chemotaxis, adhesion and activation, increase in vascular permeability, smooth muscle cell contraction and modulation of cell proliferation (reviewed in [9]), have justified the interest in their biosynthesis and role in inflammatory skin diseases. A number of reports have described the presence of leukotriene B_4-like material in psoriatic lesions [13–16], which raises the question whether keratinocytes are an effective source of leukotrienes. Grabbe et al. [17] reported the release of leukotriene B_4-like material from keratinocytes based on analysis by high-pressure liquid chromatography (HPLC), radioimmunoassay, bioassay and inhibition by the 5-lipoxygenase inhibitor BW755C. Nevertheless, the concentration of the 5-lipoxygenase inhibitor used was too high to rule out nonspecific effects, and the immunoreactive and bioactive material from HPLC fractions were too wide and not consistent with the HPLC profile obtained with leukotriene B_4 standard. Other laboratories including ours have not been able to find leukotrienes derived from arachidonic acid in incubations of epidermal cell suspensions [18–20].

Dual transfection studies have demonstrated that both 5-lipoxygenase and FLAP are necessary for cellular leukotriene synthesis [21]. The expression of 5-lipoxygenase and FLAP is essentially restricted to cells of the myeloid lineage. The concept that the cell's ability to produce leukotrienes depends on their state of differentiation has been extensively demonstrated in the HL-60 cell line. Various groups have reported that undifferentiated HL-60 cells produced only small amounts of leukotrienes compared with differentiated granulocytic or monocytic cells [22–25]. During differentiation of HL-60 cells, a simultaneous induction of both 5-lipoxygenase and FLAP has been observed, which correlates with the increase in leukotriene biosynthetic ability [26]. Janssen-Timmen et al. [27] demonstrated upregulated production of 5-HETE and leukotriene B_4 and the expression of 5-lipoxygenase protein and messenger RNA (mRNA) in differentiated HaCaT cells, a transformed human keratinocyte cell line. They also stated in this work that primary human keratinocytes express 5-lipoxygenase mRNA, but these data were not presented. Breton et al. [20] recently evaluated the capacity of human keratinocytes to synthesise 5-lipoxygenase-derived products by determining the expression of protein and mRNA for 5-lipoxygenase and FLAP in subcellular fractions. Results using Western-blot and reverse transcriptase-polymerase chain reaction (RT-PCR) techniques showed that neither bands corresponding to 5-lipoxygenase or FLAP proteins nor mRNA for 5-lipoxygenase and FLAP were detectable in all the fractions examined [20]. It therefore seems clear that 5-lipoxygenase of infiltrating leucocytes is required for effective leukotriene synthesis by keratinocytes.

Keratinocyte transcellular biosynthesis of leukotrienes

The primary product of leukotriene biosynthesis, leukotriene A_4, is not only an intracellular intermediate in the synthesis of leukotrienes B_4 and C_4 but may also be released from activated leucocytes [28, 29]. Leukotriene A_4 can then be taken up by adjacent cells devoid of 5-lipoxygenase activity but expressing leukotriene A_4 hydrolase and/or leukotriene C_4 synthase. Both enzymes are much more widely distributed than 5-lipoxygenase. This process termed transcellular metabolism has been observed in a wide range of cell types [30–37], and it allows amplification of leukotriene production in inflamed tissue.

We [18] and Iversen et al. [19] observed production of leukotriene B_4 by transcellular metabolism in mixed incubations of neutrophils with fresh human epidermal cells or human keratinocytes in culture, respectively. Like the leucocyte leukotriene A_4 hydrolase [38–40], epidermal leukotriene A_4 hydrolase is located in the cytosolic fraction [41, 42] and exhibits kinetic and physicochemical properties similar to the leucocyte enzyme [43]. A glutathione-S-transferase-dependent conjugation of leukotriene A_4 methyl ester to leukotriene C_4 methyl ester has been observed primarily in human and rodent skin [44]. The synthesis of peptidoleukotrienes by human epidermis from exogenous leukotriene A_4 was definitively demonstrated by Iversen et al. [45].

Keratinocyte metabolism of leukotrienes B_4 and C_4

Leukotrienes B_4 and C_4 are transported out of the cell by two different energy-dependent carrier systems. A multidrug resistance-associated protein has been identified as the carrier system mediating an ATP-dependent transport of leukotriene C_4 and structurally related conjugates (reviewed in [9, 46]).

Leukotriene C_4 is converted to leukotriene D_4 through cleavage of glutamic acid by γ-glutamyl transferase, which is widely distributed on cell surfaces and detectable in plasma. A variety of peptidases catalyse the further metabolism of leukotriene D_4 to leukotriene E_4 (reviewed in [46, 47]).

Degradation of leukotrienes B_4 and E_4 results in their biological inactivation. Degradation starts with their ω-oxidation, yielding ω-hydroxy-, ω-aldehyde- and ω-carboxy-leukotrienes [46, 48–52]. Following activation by acyl coenzyme A synthase, ω-carboxy-leukotrienes are subjected to β-oxidation from the ω-end predominantly in peroxisomes [51].

In keratinocytes, alternative pathways for leukotriene B_4 metabolism, which starts with dehydrogenation at C12, have been described [53]. Leukotriene B_4-12-hydroxy dehydrogenase has recently been cloned and a high expression of this enzyme has been found in kidney, liver and intestine but not in leucocytes. No data concerning expression on skin were provided [54]. The 12-oxo-leukotriene B_4 generated is either reduced to 10, 11-dihydro products (this pathway was previously described in porcine leucocytes [55]) which may subsequently undergo β- or ω-oxidation or chain elongation, or its conjugation at C6 with glutathione by glutathione-S-transferase. A conventional glutathione-S-transferase appears to be involved rather than a highly specific enzyme such as leukotriene C_4 synthase. The product is reduced at C12, yielding 5,12-dihydroxy-6-S-glutathiolyl-eicosa-7,9,14-trienoic acid, a structural analogue of leukotriene C_4 termed c-leukotriene B_3, which can similarly be converted to d-leukotriene B_3 and e-leukotriene B_3, which are structural analogues of leukotrienes D_4 and E_4, respectively [53]. The biologic functions of these compounds remain to be established.

Primary oxygen insertion in the arachidonic acid molecule at C11 and C15: the PGHS pathway

PGHS enzymes

PGHS catalyses the first step in the formation of prostaglandins and thromboxanes. The enzyme is also bifunctional, with the two enzymatic activities – cyclooxygenase and peroxidase – located in different sites of the molecule. The cyclooxygenase active site catalyses the conversion of arachidonic acid into the endoperoxide prostaglandin G_2, whereas the peroxidase active site catalyses the reduction of the C15 hydroperoxy group of prostaglandin G_2 to the corresponding alcohol, forming prostaglandin H_2. Newly formed prostaglandin H_2 is subsequently converted into what are considered the biologically active prostanoids, prostaglandins D_2, E_2, $F_{2\alpha}$ and I_2, or thromboxane A_2 (reviewed in [4, 56]).

Two PGHS isozymes encoded by two different genes termed PGHS-1 and PGHS-2 have been described. Both enzymes are located in the endoplasmic reticulum and nuclear envelope. PGHS-1 is an enzyme present in almost all mammalian cells, and its main function appears to be the maintenance of the prostaglandin production rate necessary for normal cellular processes as a "housekeeping" enzyme. PGHS-2 is undetectable in most mammalian tissues, but the expression of this isoenzyme can be rapidly and transiently induced by cytokines, growth factors and tumour promoters (reviewed in [56–58]). Primary structures of PGHS-1 and PGHS-2 are 60% identical. The kinetic properties of the two enzymes are quite similar; but despite the overall similarities between PGHS-1 and PGHS-2, it should be noted that there are subtle differences between the active sites of PGHS-1 and PGHS-2, as evidenced by their different affinities towards some fatty acid substrates [59] and nonsteroidal antiinflammatory drugs [60, 61].

Keratinocyte PGHS-pathway

There is some controversy in the literature regarding the profile of prostanoid synthesis by keratinocytes. This may be related to the use of freshly isolated vs. cultured cells, species differences, methods employed to obtain, handle and stimulate the cells or the different analytical techniques used. Arachidonic acid metabolism has been characterized in fresh human epidermal cell suspensions [18, 62, 63] and also in human as well as animal cultured keratinocytes, with major emphasis on primary cultures of mouse or human cells [64, 65]. Several reports have shown that fresh human epidermal cells in suspension produce prostaglandin E_2 as the main prostanoid, but prostaglandins $F_{2\alpha}$ and D_2 are also produced [18, 62, 63]. Most investigators found that the predominant PGHS products of keratinocytes in culture were prostaglandins E_2 and $F_{2\alpha}$ [66–68]. There is lack of agreement as to whether prostaglandins I_2 and D_2, and thromboxane A_2, are synthesised by keratinocytes or from cells other than keratinocytes that usually contaminate the cultures prepared from whole epidermis [69]. 6-Ketoprostaglandin $F_{1\alpha}$, the stable metabolite of prostaglandin I_2, was detected in preconfluent cultures of a rat keratinocyte line free of contaminating fibroblasts [65] and in primary cultures of human keratinocytes, in both preconfluent and confluent states [66]. These data suggest that keratinocytes in culture may be capable of making small amounts of prostaglandin I_2.

Prostanoids act as short-lived autocrine and paracrine mediators that induce a wide spectrum of biological activities, including cell proliferation, smooth muscle contraction, platelet homeostasis and immunomodulation (reviewed in [70, 71]). Numerous reports indicate the involvement of prostaglandins in the process of wound repair in the skin, in vasodilatation associated with inflammation and in the erythema induced by ultraviolet (UV) light.

Since 1967 when Miller reported a significant suppression of UVB-induced erythema by administration of aspirin [72], there have been many reports describing the integral role of prostaglandins in mediating UVB-induced erythema [73–76]. It has been stated that prostaglandins are also involved as mediators in the early stages of the UVA-induced inflammatory responses [77]. *In vitro* UVA and UVB radiations are reported to be inducers of human keratinocyte prostanoid synthesis and release [78, 79]. There is evidence that UVB exposure induces the production of keratinocyte-derived cytokines, including interleukin-1α (IL-1α) [67]. IL-1 increases prostaglandin synthesis in keratinocytes, which are themselves a source of the cytokine. Grewe et al. [80] have shown the involvement of endogenously produced

IL-1 and tumour necrosis factor α (TNFα) in UVB-induced prostaglandin E_2 release by transformed human keratinocytes. IL-1 has been found to act via the type 1 IL-1 receptor and to account for the major amount of prostaglandin E_2 released, whereas TNFα acts via the 55-kDa TNF receptor, contributing with a minor effect. Grewe et al. [80] found that the increasing release of prostaglandin E_2 was partly due to upregulation of PGHS activity of keratinocytes, which was mediated in an autocrine manner by the combined action of IL-1 plus TNFα.

Role of the PGHS pathway on keratinocyte proliferation and differentiation

Keratinocyte proliferation and differentiation are complex processes that are influenced by prostanoids. The participation of prostaglandin synthesis in keratinocyte proliferation has been suggested in several ways. It has been reported that phorbol ester-induced hyperproliferation in mouse epidermis is associated with an increase in endogenous synthesis of prostaglandin E_2 [81]; indomethacin blocks the effect of phorbol ester, and this effect is restored by topical application of prostaglandin E_2 [82, 83]. Prostaglandins have been shown to modulate cultured keratinocyte proliferation either through endogenous production of prostaglandin E_2 [66] or by addition of prostaglandin E_2 to cultured cells [84]. Synthesis and response to prostaglandin E_2 appears to be related to the state of growth. Prostaglandin E_2 is a growth-promoting autacoid in nonconfluent keratinocyte cultures. Confluent cultures produce less and are unresponsive to the prostaglandin. Pentland and Needleman [66] showed that the increase in endogenous prostaglandin E_2 accumulation observed in nonconfluent keratinocyte cultures is apparently mediated by an increase in PGHS activity.

It has been shown that the phorbol ester-induced hyperproliferation of epidermis correlates with the expression of PGHS-2 [85, 86]. Normal epidermis of adult mice showed constitutive expression of PGHS-1, while no expression of PGHS-2 was detectable [85]. When treated with phorbol ester, a rapid and transient increase of PGHS-2 mRNA and protein occurred, while PGHS-1 remained unchanged. This finding correlated with the increased tissue levels of prostaglandins E_2 and $F_{2\alpha}$, which were required for development of epidermal hyperplasia and wound healing. In contrast to the stimulating effect of a single phorbol ester application on epidermal PGHS-2 mRNA and protein, repeated application of the tumour promoter under experimental conditions leading to chronic hyperplasia did not cause accumulation of PGHS-2 mRNA or protein [85, 86]. These data were further supported by Maldve and Fischer [87], who showed the involvement of protein kinase C in PGHS-2 expression induced by phorbol esters in murine keratinocytes. The expression of PGHS-2 was induced in primary keratinocyte cultures by both phorbol ester and non-phorbol ester tumour promoters as well as by epidermal growth factor (EGF), indicating that there are different signalling pathways involved in the upregulation of PGHS-2 [87]. Interestingly, it has been found that retinoids suppress the phorbol ester-induced increase in PGHS-2 expression in the squamous carcinoma cell line 1483 [88].

It has been shown that tumorigenic and nontumorigenic keratinocytes in culture constitutively express both PGHS-1 and PGHS-2 at different ratios. Benign and in particular malignant epidermal tumours constitutively overexpressed PGHS-2 mRNA and protein, whereas the expression of PGHS-1 is similar to that found in healthy tissue [85, 86]. This fact, together with the anti-tumour-promoting effect of indomethacin, suggests that PGHS-2 plays a critical role in skin carcinogenesis, and it could thus be expected that the promotion of skin tumours will be particularly sensitive to drugs that selectively inhibit the activity of PGHS-2.

In addition to prostanoid synthesis, PGHS also catalyses the formation of 13- and 9-hydroperoxyoctadecadienoic (HPODE) acids from linoleic acid. The corresponding hydroxyoctadecadienoic (HODE) acids represent the major oxygenated metabolites of linoleic acid produced by cells [89]. HPODEs and HODEs have been found in high amounts in psoriatic lesions [90–92], and the reported biological activities suggest that they are involved in inflammatory and proliferative responses [93–96]. There is growing evidence that HPODEs and HODEs play a role as elements of signal pathways necessary for cell mitogenesis when stimulated by growth factors or oncogenic transformation. In particular, linoleic acid, HPODEs and HODEs enhance the response of epithelial cells and fibroblasts to insulin and EGF, respectively [97–99]. The proliferative effect of free linoleic acid-derived hydroperoxides was mediated by activation of a protein kinase in response to mitogens and c-*fos*, c-*jun* and c-*myc* mRNA expression [100].

We observed that PGHS, and mainly PGHS-2, was the enzyme responsible for the synthesis of both 13-HODE and 9-HODE in endothelial cells and dermal fibroblasts, rather than 15-lipoxygenase [101, 102]. Loftin and Eling [103] have recently reported an identical role for PGHS-2 in the EGF-stimulated mouse keratinocytes BALB/MK. They found that EGF induced the expression of PGHS-2, resulting in increased metabolism of arachidonic acid into prostaglandins E_2 and $F_{2\alpha}$ and linoleic acid into 13-HODE and 9-HODE. Inhibition of PGHS by indomethacin reduced EGF-induced DNA synthesis and cellular proliferation. Furthermore, changes in PGHS-2 expression in response to EGF, insulin and dexamethasone correlated with the mitogenic effects of these agents. Nevertheless, they did not establish the role of specific eicosanoids or octadecanoids in the regulation of BALB/MK proliferation. The fact that linoleic acid is a substantially better substrate for PGHS-2 than for PGHS-1 also supports the role of PGHS-2 in the biosynthesis of HPODEs and HODEs [59, 101]. Although the presence of 15-lipoxygenase has been reported in human epidermal cells [104], PGHS might be relevant in the synthesis of 9-HODE, 13-HODE [105] and 15-HETE. PGHS-2 could be particularly significant in stimulated keratinocytes, as seen in other skin cells [101, 102, 106, 107].

The high levels of PGHS-1 found in highly differentiated and specialized cell types such as endothelium, renal collecting tubules, macrophages or platelets suggest that PGHS-1 is related with the degree of differentiation. There is a growing body of evidence to support this concept (reviewed in [56–58]).

PGHS-2 is expressed in few tissues in the absence of inflammatory or other stimuli. Notable exceptions are the brain and the macula densa [56–58]. The fact that transgenic mice that did not express PGHS-2 developed severe nephropathy that limited their life span [108] suggests that, unlike PGHS-1, PGHS-2 is required for proper mammalian development.

Different studies have revealed a correlation between the state of differentiation of epidermal cells with PGHS activity, and the disruption of normal arachidonic acid metabolite levels led to altered growth and differentiation of skin components. Cameron et al. [68], investigating prostaglandin synthesis in density-fractionated mouse epidermal cells, observed that fractions containing the more differentiated cells metabolized significantly more arachidonic acid than did the fractions containing the less differentiated and highly proliferative cells. It has also been reported that endogenous prostaglandin E_2 enhanced the calcium-induced differentiation of human keratinocytes [109]. Leong et al. [110] studied the expression of PGHS-1 and PGHS-2 by immunohistochemistry in human and mouse skin biopsy sections. They observed that in normal human epidermis PGHS-2 immunostaining increased in the more differentiated suprabasal keratinocytes. Moreover, in human keratinocyte cultures, cell differentiation led to an increased expression of both PGHS-2 protein and mRNA, which suggests that in human epidermis the expression of PGHS-2 occurs as a part of normal keratinocyte differentiation [110]. Nevertheless, these results obtained from human epidermis were not consistent with those previously obtained from mouse epidermis [85]. More data are needed to understand the role (if any) of PGHS-2 in normal keratinocyte proliferation and differentiation.

Primary oxygen insertion in the arachidonic acid molecule at C12: the 12-lipoxygenase pathway

Keratinocyte 12-lipoxygenases

This pathway represents the major arachidonic acid oxygenation mechanism in epidermal cells, with total product formation generally exceeding PGHS activity [18, 63]. It has been shown that 12-HETE is the main eicosanoid produced by preparations of fresh human epidermis and the most abundant eicosanoid found in psoriatic lesions [18, 90, 111, 112].

12-Lipoxygenase catalyses the stereoselective insertion of a molecule of oxygen into arachidonic acid to yield 12(S)-HPETE. 12(S)-HPETE is in turn reduced to 12(S)-HETE by a glutathione-dependent peroxidase(s) (reviewed in [2, 3]). It has been observed that 12(*S*)-HETE is the predominant enantiomer produced by normal epidermal cell suspensions ($S/R \cong 5$) [63] and small fragments or homogenates of epidermis incubated with exogenous arachidonic acid ($S/R \cong 3$; R. Antón et al., unpublished results). These $12S/R$ ratios are lower than those found in incubations of human platelet suspensions with arachidonic acid (average $12S/R$ ratio of 70.4; R. Antón et al., unpublished results). In contrast, 12(*R*)-HETE was the main enantiomer found in psoriatic lesions [113]. Moreover, when psoriatic scales were incubated with exogenous arachidonic acid, they produced mainly 12(*R*)-HETE ($12S/R$ ratio of 0.175) [92].

Although both enantiomers of 12-HETE share several biologic activities, there are various differences in their activity profiles. Both have been reported to be capable of producing a dose-related induction of erythema and increased blood flow after topical application in human skin [114]. However, the 12(*R*)-HETE enantiomer appeared to be the most potent inducer of cutaneous inflammation [115], and only the *R* isomer caused remarkable neutrophil infiltration either into epidermis or dermis [114, 116]. It has been reported that racemic 12-HETE stimulates epidermal proliferation in some species such as guinea pig, but not in human neonatal keratinocytes [117] or in human skin organ cultures [118]. High-affinity binding sites of 12-HETE have been described in both keratinocytes and Langerhans cells [119], suggesting that 12-HETE actions are mediated by specific receptors whose number seems to be decreased in psoriasis [120].

Several groups orientated their efforts to assessing the enzymatic origin of the keratinocyte-derived 12-HETE in keratinocytes. An alternative pathway to yield 12-HETE is P-450 monooxygenase, which directly affords a mixture of 12(*R/S*)-HETE without a hydroperoxide intermediate, and the R isomer usually predominates [121]. Holtzman et al. [63] claimed a novel cytochrome P-450 with S stereopreference plays a role as a source of 12-HETE in epidermal cells. This was supported by several findings: (i) the activity was located predominantly in the particulate cell fractions; (ii) it was increased twofold by supplementation with nicotinamide-adenine dinucleotide phosphate (NADPH); (iii) 76–78% inhibition with carbon monoxide was observed; and (iv) two epoxyeicosatrienoic acids derived from arachidonic acid were identified in the preparations. Since the ratio of $12S/R$ was increased threefold in microsomal preparations in relation to the mitochondrial preparations, the authors suggested two different monooxygenase enzyme systems. However, the reason for the predominant 12(*R*)-HETE found in the psoriatic lesions [113] was not explained by these findings.

It has been established that, unlike 5- and 15-lipoxygenases, 12-lipoxygenase exists as different isoforms within the same species (reviewed in [122]). Biochemical, immunological and genetic evidence indicates that there are two major forms of 12-lipoxygenase, the "platelet"-type and the "leucocyte"-type. 12-Lipoxygenase has been detected in nonhematopoietic cell tissues, including epithelium, pituitary cell, adrenal glomerulosa, pancreatic islets, and in neurons and perineural cells. These

12-lipoxygenases differ in their immunoreactivities, substrate specificities, product profiles and cellular location. Molecular cloning from complementary DNA (cDNA) libraries generated from leucocytes, epithelial cells and reticulocytes indicated that the members of the family of 12-lipoxygenase and 15-lipoxygenase are highly homologous with each other (reviewed in [2, 122]). The features distinguishing the two 12-lipoxygenase forms are (i) high reactivity with C18 fatty acids such as linoleic acid for the leucocyte type in contrast with the low reactivity observed for the platelet-type 12-lipoxygenase, and (ii) the platelet type yields almost exclusively 12-HPETE, whereas the leucocyte type yields a mixture of 12-HPETE and 15-HPETE (3:1, respectively, for murine enzymes) [123]. It should be noted that oxygenation catalysed by both platelet-type and leucocyte-type 12-lipoxygenase also yields minor amounts of 12(R)-HPETE [124].

Takahashi et al. [125] described platelet-type 12-lipoxygenase activity in human epidermal cells based on the absence of stimulation by NADPH, inhibition with nordihydroguaiaretic acid, detection of the 12-HPETE, localization either in the cytosol or the particulate fractions (mitochondrias and microsomes), immunoprecipitation with antibodies against human platelet 12-lipoxygenase and specificity for substrate. From the data of Takahashi et al., a contribution of monooxygenases in the synthesis of 12-HETE cannot be ruled out, since these authors worked with the cytosol and with the solubilized enzymes from the particulate fractions. Nevertheless, they stated that the activity present in the fraction, which was not soluble in Tween 20, also produced 12-HPETE, which could not come from cytochrome P-450 activity, although these data were not presented.

Results obtained by Takahashi et al. [125] have been confirmed by several authors who found that platelet-type 12-lipoxygenase is overexpressed in germinal layer keratinocytes [126] and in skin tumours [127]. Among the 12-lipoxygenases, the platelet type has been found to be the predominant isoenzyme expressed in human and murine skin epidermis [125–127]. Despite the fact that leucocyte-type 12-lipoxygenase has also been found in epidermal tumour samples and phorbol ester-treated epidermis, it probably comes from the inflammatory infiltrate and/or resident nonepithelial cells [127].

Van Dijk et al. [128] described the genomic sequence of a skin-expressed murine gene closely related to murine platelet-type and leucocyte-type 12-lipoxygenase, which was originally considered to represent a pseudogene [123]. Two different laboratories have recently isolated, functionally characterized and described the cellular localization of the lipoxygenase encoded by such a gene, resulting in a novel 12-lipoxygenase [129, 130]. This epidermal 12-lipoxygenase genomic sequence was equally related to leucocyte- and platelet-type 12-lipoxygenase [129, 130], and the recombinant protein had a molecular weight of about 77 kDa, similar to that found for platelet- and leucocyte-type 12-lipoxygenase (about 75 and 77 kDa, respectively) [130]. Epidermal 12-lipoxygenase exclusively produced the S enantiomer of 12-HETE with a very low proportion of 15-HETE [129, 130]. The conversion rate of linoleic acid into 13-HODE was below 2% [130]. Epidermal-type 12-lipoxygenase thus functionally resembles platelet-type 12-lipoxygenase rather than leucocyte type.

Keratinocyte 12-lipoxygenase pathway

Although the 12-lipoxygenase pathway is the most active in epidermal cells, little attention has been paid to its products. Much interest has been given to those pathways that yield prostaglandins or leukotrienes, since they have very relevant proinflammatory activities. Hepoxilins are formed by an intramolecular rearrangement of the hydroperoxide group in 12-HPETE by transfer of the hydroxyl group to C8 or C10, and the concomitant formation of an epoxide group at C11–C12 to form the 8(R/S)-hydroxy-11(S), 12(S)-*trans*-epoxyeicosa-5Z,9E,14Z-trienoic acid (hepoxilin A_3) and 10(R/S)-hydroxy-11(S), 12(S)-*trans*-epoxyeicosa-5Z,8Z,14Z-trienoic acid (hepoxilin B_3), respectively. Hepoxilin A_3 formation can be nonenzymatically catalysed by hemin and hemoglobin, and enzymatically by a hepoxilin A_3 synthase. There is no conclusive evidence that hepoxilin B_3 can be formed enzymatically (reviewed in [6]).

We recently reported that in human epidermis 12(S)-HPETE alternatively undergoes further types of transformations [131], as occurs in other cell types and tissues [3, 6, 132]. Normal human epidermis incubated with exogenous arachidonic acid primarily produced, in addition to 12-HETE, hepoxilin A_3, hepoxilin B_3, 8,9,12-trihydroxyeicosatrienoic acid (8,9,12-THETrE) and 12-oxo-ETE via the 12-lipoxygenase pathway (see scheme in Fig. 2) [131]. The reverse phase-HPLC (RP-HPLC) fraction from incubates of small fragments of human epidermis with exogenous arachidonic acid which contained hepoxilins was composed of a single peak. This fraction, rechromatographed by straight phase-HPLC (SP-HPLC), afforded a major peak with no UV absorption between 210 and 320 nm, indicating the absence of conjugated double bonds in the molecule. The gas chromatography-mass spectrometry analysis of this compound demonstrated its hepoxilin B_3 identity.

This HPLC fraction was catalytically hydrogenated to confirm its hepoxilin B_3 identity. Racemic authentic (±)hepoxilin B_3 yielded two peaks with identical mass spectra, as expected, consistent with the 10-hydroxy-11,12-epoxy-arachidic acid structure. Unexpectedly, the corresponding epidermal hydrogenated compound yielded only one gas chromatography peak with a mass spectrum with a structure corresponding to 10-hydroxy-11,12-epoxy-arachidic acid [131]. Our results from gas chromatography-mass spectrometry analysis of the hydrogenated compounds suggested that human epidermis yields almost exclusively one of the two possible epimers of hepoxilin B_3. To confirm this point, authentic (±)hepoxilin B_3 and hepoxilin B_3 from epidermal incubates were subjected to the same extraction and purification procedures and were later derivatized with 9-anthryldiazomethane (ADAM).

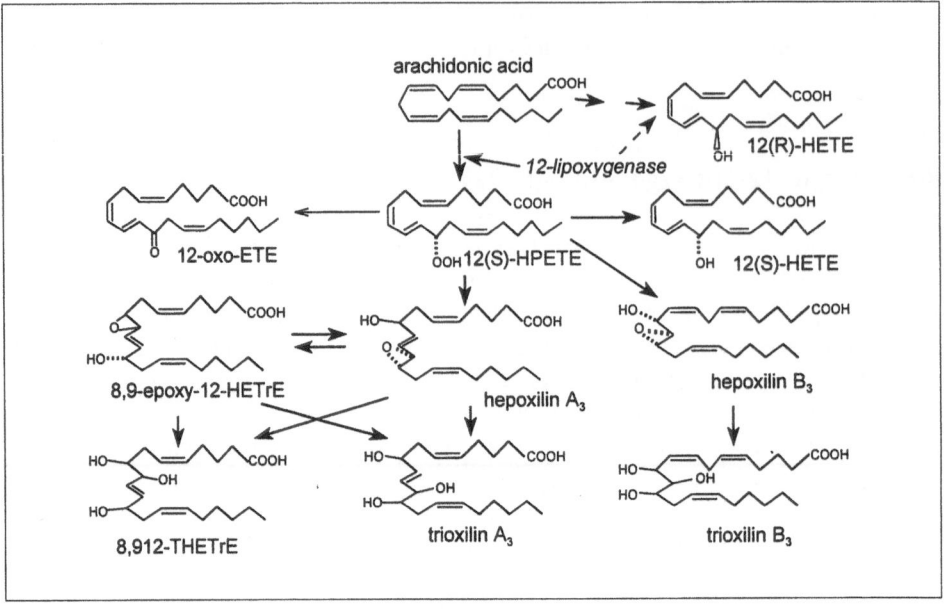

Figure 2
Diagram depicting the formation of epoxy-hydroxy and trihydroxy derivatives of arachidonic acid through the 12-lipoxygenase pathway in normal human epidermis.

The RP-HPLC of the ADAM derivatives, as shown in Figure 3, yielded a single peak in the case of hepoxilin B_3 from epidermal incubates, whereas authentic ($\pm$)hepoxilin B_3 yielded two peaks corresponding to the two 10-hydroxy epimers. These results indicated that, unlike other tissues [133–135], normal human epidermis produces only one 10-hydroxy epimer of hepoxilin B_3.

The stereoselective synthesis of hepoxilin B_3 by epidermis suggests that it may be catalysed by a specific enzyme. This concept was reinforced by the fact that incubations of 12(S)-HPETE with fresh epidermis yielded practically a single peak, whereas when incubated with boiled epidermal fragments, only a small amount of racemic hepoxilin B_3 was obtained (Fig. 4, unpublished results).

We have not directly determined the exact configuration of C10 in epidermis-derived hepoxilin B_3, but based on the chromatographic data of these compounds reported by other authors [135-137], apparently normal human epidermis produces exclusively the 10(R)-hydroxy epimer of hepoxilin B_3. Interestingly, while hepoxilin A_3 and hepoxilin B_3 are both active in enhancing bradykinin-evoked permeability in skin [138, 139], only 10(R)-hepoxilin B_3 stereospecifically enhances the vascular permeability evoked by intradermal injection of platelet-activating factor (PAF) [139].

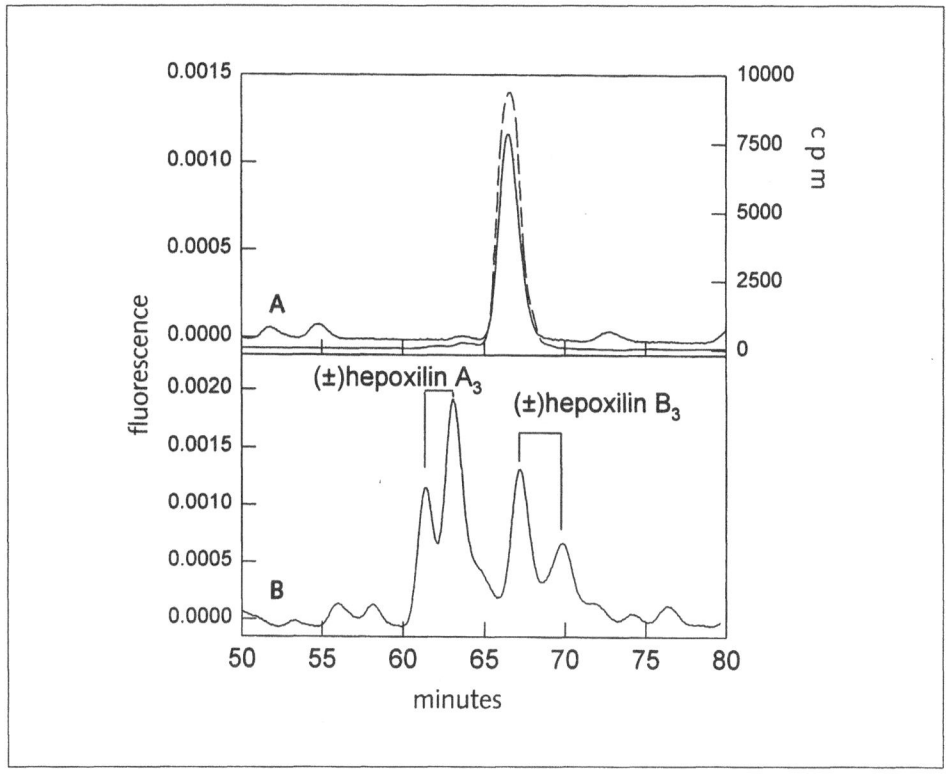

Figure 3
Reverse-phase chromatograms obtained from samples of human epidermis-derived hepox-
ilin B₃ (A), and authentic (±)hepoxilin A₃ plus (±)hepoxilin B₃ (B) derivatized with ADAM.
Epidermal hepoxilin B₃ was obtained from incubations of epidermis with [¹⁴C]-arachidonic
acid, and radioactivity was also monitored simultaneously (A, dotted line).

Coming back to the enigma of the origin of 12(R)-HETE, the 12-lipoxygenase-mediated synthesis of 12(R)-HPETE by epidermal cells is supported by the fact that formation of 12-HETE by fragments of normal epidermis incubated with exogenous arachidonic acid was not inhibited by metyrapone or clotrimazole (P-450 inhibitors), whereas the presence of esculetin (an inhibitor of 12-lipoxygenase) concentration-dependently inhibited 12-HETE without altering the ratio of 12(S)-HETE to 12(R)-HETE (R. Antón et al., unpublished results). This was also consistent with the data reported by Van Waume et al. [140], who observed that rat epidermal microsomes incubated in the presence of arachidonic acid and NADPH produced 12-hydroxy-5,8,14-eicosatrienoic acid (12-HETrE) with an S/R ratio of 65:35 and that its formation was apparently mediated by a lipoxygenase rather than

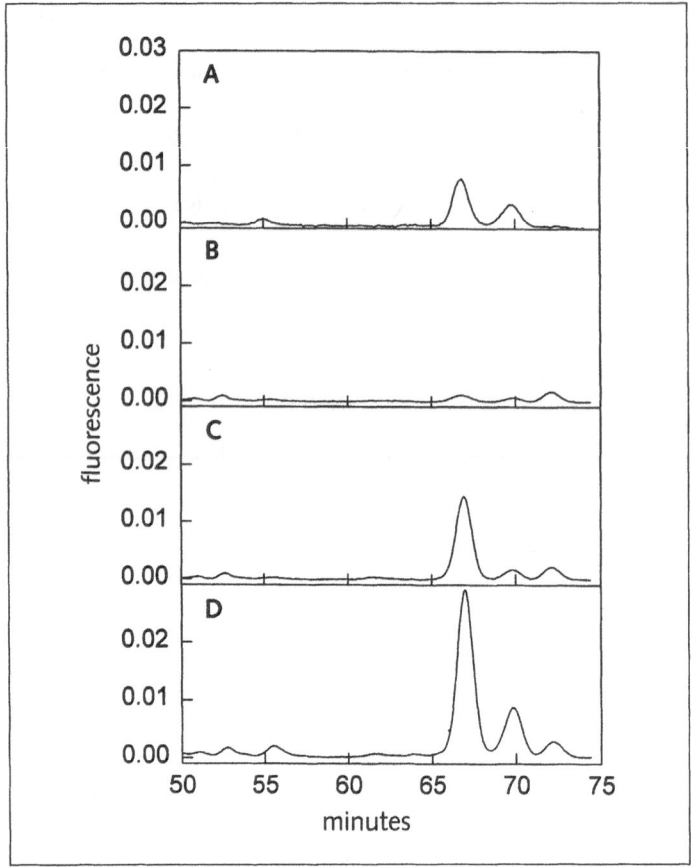

Figure 4
Chromatograms corresponding to a sample of authentic (±)hepoxilin B₃ (A), incubations of
12(S)-HPETE with boiled human epidermis (B) and with normal human epidermis (C), and
sample C plus authentic (±)hepoxilin B₃ (D).

by cytochrome P-450 [140]. Although potent proinflammatory activities of 12(R)-HETrE have been reported [141–143], at present there is no information about the occurrence of 12-HETrE *in vivo* in mammalian skin.

Assuming a 12-lipoxygenase origin in human epidermis, another possibility that could account for the infrequent *R/S* ratio of 12-HETE observed in incubations of epidermal cells and fragments is the stereoselective disappearance of 12(S)-HPETE, which should cause 12(R) enrichment of the 12(S/R)-HETE mixture. This hypothesis is supported by the fact that, while hemin catalyses hepoxilin formation either from 12(R)- or 12(S)-HPETE, rat pineal gland, suspected of expressing a hepoxilin-synthase enzyme, stereoselectively used 12(S)-HPETE to form hepoxilin A_3 [135]. The fact that normal epidermis selectively produced one of the two possible 10-hydroxy epimers of hepoxilin B_3 also suggests a selective substrate utilization in its synthesis. This is presently under investigation in our laboratory.

Hepoxilins are unstable and undergo hydrolysis to be converted into 8,11,12-tri-hydroxy-5,9,14-eicosatrienoic acid and 10,11,12-trihydroxy-5,8,14-eicosatrienoic acid, commonly termed trioxilin A_3 and trioxilin B_3, respectively [6]. This reaction may also be catalysed by a hepoxilin-epoxide hydrolase [144]. Alternatively, a glu-tathione-S-transferase may catalyse the conjugation of hepoxilin A_3 with glu-tathione to afford hepoxilin A_3-C, which is structurally related to leukotriene C_4. When the hepoxilin-epoxide hydrolase pathway was inhibited by trichloropropene oxide, metabolism of hepoxilin A_3 was diverted towards hepoxilin A_3-C [145, 146]. Both trioxilin A_3 and trioxilin B_3 are also found in the incubates of human epidermis with exogenous arachidonic acid [131] (see Fig. 2).

Occurrence of hepoxilins and trioxilins in psoriasis

We recently investigated the occurrence of hepoxilins and trioxilins *in vivo* under a pathophysiological situation such as psoriasis. Results are shown in Table 1 [147].

Table 1 - Levels of hepoxilins, trioxilins and 12-HETE found in normal epidermis and in psoriatic scales

Compound	Normal isolated epidermis		Psoriatic scales	
	ng/mg[1]	A/A$_{12\text{-HETE}}$[2]	ng/mg[1]	A/A$_{12\text{-HETE}}$
Hepoxilin A_3	–	nd[3]	–	nd
Hepoxilin B_3	< 0.2[4]	–	3.2 ± 2.3	–
Trioxilin A_3	–	nd	–	0.65 ± 0.23
8,9,12-THETrE	–	nd	–	0.05 ± 0.02
Trioxilin B_3	–	nd	–	0.32 ± 0.28
12-HETE	0.8 ± 0.3	1	21.3 ± 12.4	1

[1] *Mean ± SD, n = 5;* [2] *These data are semiquantitative values in terms of peak area relative to 12-HETE;* [3] *nd, not detected;* [4] *Hepoxilin B_3 was only detected in two samples of normal epidermis.*
Samples were extracted with methanol and the ^{18}O-labelled internal standards were added. Since standards for trioxilins and 8,9,12-THETrE were not available, semiquantitative evaluation of these compounds was accomplished by calculating the ratio of the areas of the corresponding gas chromatography peak to that of 12-HETE [147].

Not only were both hepoxilins and trioxilins found to be elevated in psoriasis, but they were also present in biologically active amounts. Figure 5 shows typical m/z 273 selected ion chromatograms for monitoring hepoxilin B_3 (hydrogenated) obtained from pooled psoriatic scales, pooled fragments of isolated normal epidermis and authentic (±)hepoxilin B_3. As aforementioned, hydrogenated hepoxilin B_3 from normal epidermis incubated with exogenous arachidonic acid yielded only one gas chromatography peak with a mass spectrum consistent with 10-hydroxy-11,12-epoxy-arachidic acid, which corresponds to peak 2 in Figure 5 [131]. Samples from authentic hydrogenated racemic hepoxilin B_3 yielded two peaks consistent with the same structure, corresponding to the two epimeric forms at C10 (peaks 2 and 3 in Fig. 5). Unexpectedly, samples from psoriatic patients showed three peaks (peaks 1, 2 and 3 in Fig. 5, respectively), with essentially identical mass spectra consistent with a 10-hydroxy-11,12-epoxy-arachidic acid structure.

In order to discuss the identity and the origin of the additional peak found in psoriatic samples (peak 1 Fig. 5), one can speculate that the enantiomer 10(S,R)-hydroxy-11(R),12(R)-epoxy was also present in psoriatic lesions, since 12(R)-HETE was the predominant structure [92, 113]. However, 10(S,R)-hydroxy-11(S),12(S)-epoxy- (configuration of (±)hepoxilin B_3) and 10(S,R)-hydroxy-11(R),12(R)-epoxy-configurations are enantiomers, which migrate together in a nonchiral stationary phase column, resulting in only two peaks (instead of three or four) corresponding to the two epimers at C10. Therefore, the most probable explanation for the additional peak found in psoriatic scales would be the presence of 10(R,S)-hydroxy-11(R),12(S)-epoxy- or/and 10(R,S)-hydroxy-11(S),12(R)-epoxy compounds in psoriatic lesions.

Monooxygenases catalyze production of cis-epoxides [11(R),12(S) or 11(S),12(R)] [148], in contrast to trans-epoxides [11(R),12(R) or 11(S),12(S)], which are produced by lipoxygenases [149]. However, it has been reported that epoxy-eicosatrienoic acids instead of hydroxy-epoxy acids are formed by the action of monooxygenases [150]. It could thus be possible that 10-hydroxy-cis-epoxides were formed by rearrangement of an arachidonic acid-peroxyl radical intermediate as a

Figure 5

GC-MS m/z 273 selected ion chromatograms of the hydrogenated methyl ester trimethyl silyl ether derivatives of the hepoxilin-HPLC fraction from pooled psoriatic scales (A), fragments of pooled isolated normal epidermis (B) and authentic (±)hepoxilin B_3 (C) processed identically. Ion m/z 273 is specific to the hydroxyl group in C10 and corresponds to the carboxy terminal fragment of hepoxilin B_3. The ordinate represents the percentage of the MS detector signal relative to the highest peak, and the abscissa represents the period of time after the sample injection. Peaks 1, 2 and 3 had essentially identical mass spectra consistent with 10-hydroxy-11,12-epoxy-arachidic acid structure; peaks 4 and 5 corresponded to the hydrogenolysis products 10,11-dihydroxy-arachidic acid.

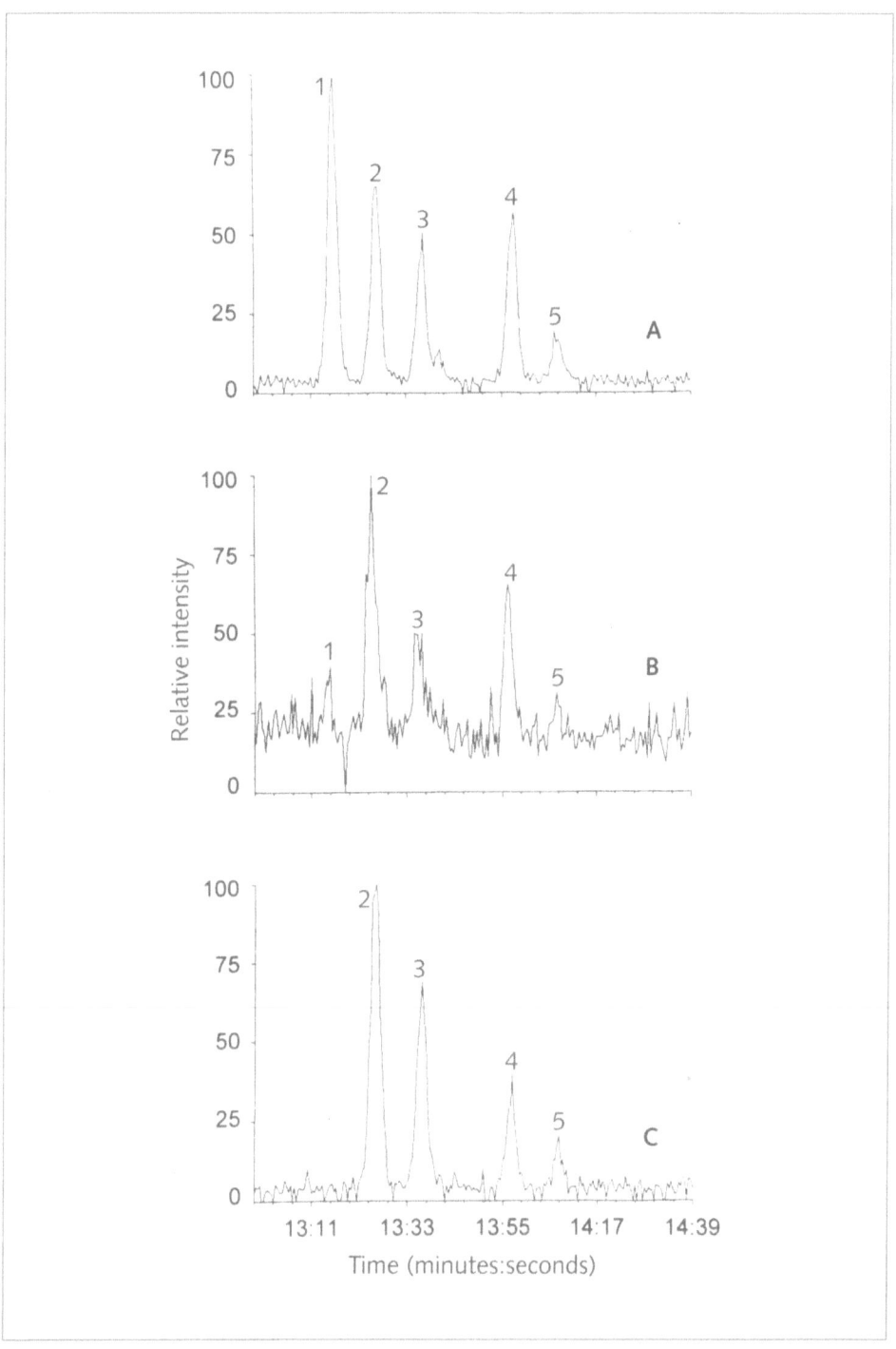

result of an autocatalytic process induced by free radicals (OH·) or by ONOO⁻ formed by the reaction of NO· with O_2·⁻ [151, 152]. This is consistent with the overexpression of inducible NO-synthase reported in psoriatic skin [153, 154]. Hepoxilin A_3 was not detected possibly due to its instability, but its formation was assured by the presence of its stable derivative, trioxilin A_3.

Support for the potential role of hepoxilins in the pathogenesis of inflammatory skin diseases, in particular in psoriasis, includes their potent action on plasma permeability when injected subcutaneously [138], release of arachidonic acid and diacylglycerol [154] in human neutrophils, and the detection of considerable amounts in psoriatic lesions [147]. Hepoxilins also cause specific receptor-dependent [155, 156] induction of Ca^{2+} mobilization from endogenous sources in neutrophils [157, 158]. As a result, Ca^{2+}-dependent phospholipase A_2 could be activated, and the translocation of 5-lipoxygenase would lead to the release of other mediators such as leukotrienes. It has been observed that a Ca^{2+} pulse in epidermal cells elicited by IL-8 promotes their proliferation [159]. The effect of hepoxilin A_3 on Ca^{2+} mobilization has been demonstrated in neutrophils, and whether or not this effect also occurs in epidermal cells should be the subject of further investigations. In any case, hepoxilins could play an autocrine role as intracellular messengers and a paracrine role by modulating leucocyte activation. The literature available on the physiological role of hepoxilin B_3 is limited. However, this compound has been shown to possess insulin secretagogue activity [160, 161], to play a possible role as a second messenger for presynaptic inhibition in aplysia sensory cells [162] and to enhance the vascular permeability evoked by intradermal injection of bradykinin and PAF [139].

Despite the fact that the biological role of hepoxilins remains to be fully elucidated (reviewed in [163]), the presence of biologically active amounts of hepoxilins and trioxilins in psoriatic lesions suggests that these compounds could play a role as modulators of the inflammatory response on skin.

Acknowledgments
Original work in the authors' laboratory was supported by grants from Institut de Recerca of the HSCSP, DGICYT grants PM88-0113 and PM92-0183 and FIS grant FIS94/1559.

References

1   Corey EJ, Albright JO, Barton AE, Hashimoto S (1980) Chemical and enzymatic synthesis of 5-HPETE, a key biological precursor of slow-reacting substance of anaphylaxis (SRS) and 5-HETE. *J Am Chem Soc* 102: 1435–1436
2   Yamamoto S (1992) Mammalian lipoxygenases: molecular structures and functions. *Biochim Biophys Acta* 1128: 117–131

3   Pace-Asciak CR, Asotra S (1989) Biosynthesis, catabolism and biological properties of HPETEs, hydroperoxide derivatives of arachidonic acid. *Free Radic Biol Med* 7: 409–433

4   Porter NA (1980) Prostaglandin endoperoxides. In: WA Pryor (ed): *Free radicals in biology*, vol 4. Academic Press, London, 261–294

5   Serhan CN, Romano M (1995) Lipoxin biosynthesis and actions: role of the human platelet LX-synthase. *J Lipid Mediat Cell Signal* 12: 293–306

6   Pace-Asciak CR, Reynaud D, Demin PM (1995) Hepoxilins: a review on their enzymatic formation, metabolism and chemical synthesis. *Lipids* 30: 107–114

7   Musser JH, Kreft AF (1992) 5-Lipoxygenase: properties, pharmacology and the quinolinyl(bridged)aryl class of inhibitors. *J Med Chem* 35: 2501–2524

8   Ford-Hutchinson AW, Gresser M, Young RN (1994) 5-Lipoxygenase. *Annu Rev Biochem* 63: 383–417

9   Denzlinger C (1996) Biology and pathophysiology of leukotrienes. *Crit Rev Oncol Hematol* 23: 167–223

10 Hill E, Maclouf J, Murphy RC, Henson PM (1992) Reversible membrane association of neutrophil 5-lipoxygenase is accompanied by retention of activity and a change in substrate specificity. *J Biol Chem* 207: 22048–22053

11 Claesson H-E, Haeggström JZ, Odlander B, Medina JF, Wetterholm A, Jakobsson P-J, Radmark O (1991) The role of leukotriene A_4 hydrolase in cells and tissues lacking 5-lipoxygenase. *Adv Exp Med Biol* 314: 307–315

12 Lam BK, Penrose JF, Xu K, Austen KF (1995) Leukotriene C_4 synthase. *J Lipid Mediat Cell Signal* 12: 333–341

13 Brain SD, Camp RDR, Dowd PM, Black KA, Woollard PM, Mallet AI, Greaves MW (1982) Psoriasis and leukotriene B_4. *Lancet* 2: 762–763

14 Brain S, Camp R, Dowd P, Black KA, Greaves M (1984) The release of leukotriene B_4-like material in biologically active amounts from the lesional skin of patients with psoriasis. *J Invest Dermatol* 83: 70–73

15 Brain SD, Camp RDR, Cunningham FM, Dowd PM, Greaves MW, Black KA (1984) Leukotriene B_4-like material in scale of psoriatic lesions. *Br J Pharmacol* 83: 313–317

16 Ruzicka T, Simmet T, Peskar BA, Ring J (1986) Skin levels of arachidonic acid-derived inflammatory mediators and histamine in atopic dermatitis and psoriasis. *J Invest Dermatol* 86: 105–108

17 Grabbe J, Rosenbach T, Czarnetzki BM (1985) Production of LTB_4-like chemotactic arachidonate metabolites from human keratinocytes. *J Invest Dermatol* 85: 527–530

18 Solá J, Godessart N, Vila L, Puig L, De Moragas JM (1992) Epidermal cell-polymorphonuclear leukocyte cooperation in the formation of leukotriene B_4 by transcellular biosynthesis. *J Invest Dermatol* 98: 333–339

19 Iversen L, Fogh K, Ziboh VA, Kristensen P, Schmedes A, Kragballe K (1993) Leukotriene B_4 formation during human neutrophil keratinocyte interactions: evidence for transformation of leukotriene A_4 by putative keratinocyte leukotriene A_4 hydrolase. *J Invest Dermatol* 100: 293–298

20 Breton J, Woolf D, Young P, Chabot-Fletcher M (1996) Human keratinocytes lack the components to produce leukotriene B_4. *J Invest Dermatol* 106: 162–167

21 Ford-Hutchinson AW (1991) FLAP: a novel drug target for inhibiting the synthesis of leukotrienes. *Trends Pharmacol Sci* 12: 68–70

22 Ziboh VA, Wong T, Wu MC, Yunis AA (1986) Lipoxygenation of arachidonic acid by differentiated and undifferentiated human promyelocytic HL-60 cells. *J Lab Clin Med* 108: 161–166

23 Goerig M, Habenicht AJR, Zeh W, Salbach P, Kommerell B, Rothe DER, Nastainczyk W, Glomset JA (1988) Evidence for coordinate, selective regulation of eicosanoid synthesis in platelet-derived growth factor-stimulated 3T3 fibroblasts and in HL-60 cells induced to differentiate into macrophages or neutrophils. *J Biol Chem* 263: 19384–19391

24 Kargman S, Rouzer CA (1989) Studies on the regulation, biosynthesis and activation of 5-lipoxygenase in differentiated HL60 cells. *J Biol Chem* 264: 13313–13320

25 Bennett CF, Chiang MY, Monia BP, Crooke ST (1993) Regulation of 5-lipoxygenase and 5-lipoxygenase-activating protein expression in HL-60 cells. *Biochem J* 289: 33–39

26 Reid GK, Kargman S, Vickers PJ, Mancini JA, Leveille C, Ethier D, Miller DK, Gillard JW, Dixon RAF, Evans JF (1990) Correlation between expression of 5-lipoxygenase-activating protein, 5-lipoxygenase and cellular leukotriene synthesis. *J Biol Chem* 265: 19818–19823

27 Janssen-Timmen U, Vickers PJ, Wittig U, Lehmann WD, Stark H-J, Fusenig NE, Rosenbach T, Radmark O, Samuelsson B, Habenicht AJR (1995) Expression of 5-lipoxygenase in differentiating human skin keratinocytes. *Proc Natl Acad Sci USA* 92: 6966–6970

28 Sala A, Testa T, Folco G (1996) Leukotriene A_4, and not leukotriene B_4, is the main 5-lipoxygenase metabolite released by bovine leukocytes. *FEBS Lett* 388: 94–98

29 Sala A, Bolla M, Zarini S, Müller-Peddinghaus R, Folco G (1996) Release of leukotriene A_4 versus B_4 from human polymorphonuclear leukocytes. *J Biol Chem* 271: 17944–17948

30 Lindgren JA, Edenius C (1993) Transcellular biosynthesis of leukotrienes and lipoxins via leukotriene A_4 transfer. *Trends Pharmacol Sci* 14: 351–354

31 Marcus AJ, Hajjar DP (1993) Vascular transcellular signalling. *J Lipid Res* 34: 2017–2031

32 Brady HR, Papayianni A, Serhan CN (1994) Leukocyte adhesion promotes biosynthesis of lipoxygenase products by transcellular routes. *Kidney Int* 45: S90–S97

33 Odlander B, Jakobsson PJ, Medina JF, Radmark O, Yamaoka KA, Rosen A, Claesson HE (1989) Formation and effects of leukotriene B_4 in human lymphocytes. *Int J Tissue React* 11: 277–289

34 Jakobsson PJ, Odlander B, Claesson HE (1991) Effects of monocyte-lymphocyte interactions on the synthesis of leukotriene B_4. *Eur J Biochem* 196: 395–400

35 Maugeri N, Evangelista V, Celardo A, Dell'Elba G, Martelli N, Piccardoni P, De Gaetano G, Cerletti C (1994) Polymorphonuclear leukocyte-platelet interaction: role of

P-selectin in thromboxane and leukotriene C_4 cooperative synthesis. *Thromb Haemost* 72: 450–456

36 Feinmark SJ, Cannon PJ (1987) Vascular smooth muscle cell leukotriene C_4 synthesis: requirement for transcellular leukotriene A_4 metabolism. *Biochim Biophys Acta* 922: 125–135

37 Fukai F, Suzuki Y, Ohtaki H, Katayama T (1993) Rat hepatocytes generate peptide leukotrienes from leukotriene A_4. *Arch Biochem Biophys* 305: 378–384

38 Radmark O, Shimizu T, Jörnvall H, Samuelsson B (1984) Leukotriene A_4 hydrolase in human leukocytes. Purification and properties. *J Biol Chem* 259: 12339–12345

39 Maycock AL, Anderson MS, DeSousa DM, Kuehl FA Jr (1982) Leukotriene A_4: preparation and enzymatic conversion in a cell-free system to leukotriene B_4. *J Biol Chem* 257: 13911–13914

40 Jakschik BA, Kuo CG (1983) Characterization of leukotriene A_4 and B_4 biosynthesis. *Prostaglandins* 25: 767–782

41 Iversen L, Ziboh VA, Shimizu T, Ohishi N, Radmark O, Wetterholm A, Kragballe K (1994) Identification and subcellular localization of leukotriene A_4-hydrolase activity in human epidermis. *J Dermatol Sci* 7: 191–201

42 Ikai K, Okano H, Horiguchi Y, Sakamoto Y (1994) Leukotriene A4 hydrolase in human skin. *J Invest Dermatol* 102: 253–257

43 Iversen L, Kristensen P, Nissen JB, Merrick WC, Kragballe K (1955) Purification and characterization of leukotriene A_4 hydrolase from human epidermis. *FEBS Lett* 358: 316–322

44 Agarwal R, Raza H, Allyn DL, Bickers DR, Mukhtar H (1992) Glutathione-*S*-transferase-dependent conjugation of leukotriene A_4-methyl ester to leukotriene C4-methyl ester in mammalian skin. *Biochem Pharmacol* 44: 2047–2053

45 Iversen L, Kristensen P, Gron B, Ziboh VA, Kragballe K (1994) Human epidermis transforms exogenous leukotriene A_4 into peptide leukotrienes: possible role in transcellular metabolism. *Arch Dermatol Res* 286: 261–265

46 Keppler D (1992) Leukotrienes: biosynthesis, transport, inactivation and analysis. *Rev Physiol Biochem Pharmacol* 121: 1–30

47 König W, Schönfeld W, Raulf M, Köller J, Scheffer J, Brom J (1990) The neutrophil and leukotrienes – role in health and disease. *Eicosanoids* 3: 1–22

48 Powell WS (1984) Properties of leukotriene B_4 20-hydroxylase from polymorphonuclear leukocytes. *J Biol Chem* 259: 3082–3089

49 Soberman RJ, Harper TW, Murphy RC, Austen KF (1985) Identification and functional characterization of leukotriene B_4 20-hydroxylase of human polymorphonuclear leukocytes. *Proc Natl Acad Sci USA* 82: 2292–2295

50 Kikuta Y, Kusunose E, Kondo T, Yamamoto S, Kinoshita H, Kusunose M (1994) Cloning and expression of a novel form of leukotriene B_4 ω-hydroxylase from human liver. *FEBS Lett* 348: 70–74

51 Jedlitschky G, Huber M, Völkl A, Muller M, Leier I, Müller J, Lehmann WD, Fahimi

HD, Keppler D (1991) Peroxisomal degradation of leukotrienes by β-oxidation from the ω-end. *J Biol Chem* 266: 24763–24772

52 Huber M, Müller J, Leier I, Jedlitschky G, Ball HA, Moore KP, Taylor GW, Williams R, Keppler D (1990) Metabolism of cysteinyl leukotrienes in monkey and man. *Eur J Biochem* 194: 309–315

53 Wheelan P, Zirrolli JA, Morelli JG, Murphy RC (1993) Metabolism of leukotriene B_4 by cultured human keratinocytes. Formation of glutathione conjugates and dihydro metabolites. *J Biol Chem* 268: 25439–25448

54 Yokomizo T, Ogawa Y, Uozumi N, Kume K, Izumi T, Shimizu T (1996) cDNA cloning, expression and mutagenesis study of leukotriene B_4 12-hydroxydehydrogenase. *J Biol Chem* 271: 2844–2850

55 Powell WS, Gravelle F (1989) Metabolism of leukotriene B_4 to dihydro and dihydro-oxo products by porcine leukocytes. *J Biol Chem* 264: 5364–5369

56 Smith WL, DeWitt DL (1996) Prostaglandin endoperoxide H synthase-1 and -2. *Adv Immunol* 62: 167–215

57 Goppelt-Struebe M (1995) Regulation of prostaglandin endoperoxide synthase (cyclooxygenase) isozyme expression. *Prostaglandins Leukot Essent Fatty Acids* 52: 213–222

58 Wu KK (1996) Cyclooxygenase 2 induction: molecular mechanism and pathophysiologic roles. *J Lab Clin Med* 128: 242–245

59 Laneuville O, Breuer DK, Xu N, Huang ZH, Gage DA, Watson JT, Lagarde M, DeWitt DL, Smith WL (1995) Fatty acid substrate specificities of human prostaglandin endoperoxide H synthases-1 and -2. Formation of 12-hydroxy-(9Z,13E/Z,15Z)-octadecatrienoic acids from α-linoleic acid. *J Biol Chem* 270: 19330–19336

60 Meade EA, Smith WL, DeWitt DL (1993) Differential inhibition of prostaglandin endoperoxide synthase (cyclooxygenase) isozymes by aspirin and other non-steroidal anti-inflammatory drugs. *J Biol Chem* 268: 6610–6614

61 Masferrer JL, Zweifel BS, Manning PT, Hauser SD, Leahy KM, Smith WG, Isakson PC, Seibert K (1994) Selective inhibition of inducible cyclooxygenase 2 *in vivo* is anti-inflammatory and nonulcerogenic. *Proc Natl Acad Sci USA* 91: 3288–3232

62 Hammarström S, Lindgren JÅ, Marcelo C, Duell EA, Anderson TF, Voorhees JJ (1979) Arachidonic acid transformations in normal and psoriatic skin. *J Invest Dermatol* 73: 180–183

63 Holtzman MJ, Turk J, Pentland A (1989) A regiospecific monooxygenase with novel stereopreference is the major pathway for arachidonic acid oxygenation in isolated epidermal cells. *J Clin Invest* 84: 1446–1453

64 Kondoh H, Sato Y, Kanoh H (1985) Arachidonic acid metabolism in cultured mouse keratinocytes. *J Invest Dermatol* 85: 64–69

65 Kvedar JC, Levine L (1987) Modulation of arachidonic acid metabolism in a cultured newborn rat keratinocyte cell line. *J Invest Dermatol* 88: 124–129

66 Pentland AP, Needleman P (1986) Modulation of keratinocyte proliferation *in vitro* by endogenous prostaglandin synthesis. *J Clin Invest* 77: 246–251

67   Pentland AP, Mahoney MG (1990) Keratinocyte prostaglandin synthesis is enhanced by IL-1. *J Invest Dermatol* 94: 43–46

68   Cameron GS, Baldwin JK, Jasheway DW, Patrick KE, Fisher SM (1990) Arachidonic acid metabolism varies with the state of differentiation in density gradient-separated mouse epidermal cells. *J Invest Dermatol* 94: 292–296

69   Ujihara M, Horiguchi Y, Ikai K, Urade Y (1988) Characterization and distribution of prostaglandin D synthase in rat skin. *J Invest Dermatol* 90: 448–451

70   Needleman P, Turk J, Jakschik BA, Morrison AR, Lefkowith JB (1986) Arachidonic acid metabolism. *Ann Rev Biochem* 55: 69–102

71   Weissmann G (1993) Prostaglandins as modulators rather than mediators of inflammation. *J Lipid Mediat* 6: 275–286

72   Miller WS, Ruderman FR, Smith JG Jr (1967) Aspirin and ultraviolet light-induced erythema in man. *Arch Dermatol* 95: 357–358

73   Snyder DS (1976) Effect of topical indomethacin on UVR-induced redness and prostaglandin E levels in sunburned guinea pig skin. *Prostaglandins* 11: 631–643

74   Black AK, Fincham N, Greaves MW, Hensby CN (1980) Time course changes in levels of arachidonic acid and prostaglandins D_2, E_2 and $F_{2\alpha}$ in human skin following ultraviolet B irradiation. *Br J Clin Pharmacol* 10: 453–457

75   Ruzika T, Walter JF, Printz MP (1983) Changes in arachidonic acid metabolism in UV-irradiated hairless mouse skin. *J Invest Dermatol* 81: 300–303

76   Nagayo K (1982) Role of prostaglandin E_2 and $F_{2\alpha}$ in UVB-induced erythema. *Jap J Derm Soc* 91: 645–652

77   Imokawa G, Tejima T (1989) A possible role of prostaglandins in PUVA-induced inflammation: implication by organ cultured skin. *J Invest Dermatol* 91: 296–300

78   Hanson D, DeLeo V (1990) Long-wave ultraviolet light induces phospholipase activation in cultured human epidermal keratinocytes. *J Invest Dermatol* 95: 158–163

79   DeLeo VA, Horlick H, Hanson D, Eisenger M, Harber L (1984) Ultraviolet radiation induces changes in membrane metabolism of human keratinocytes in culture. *J Invest Dermatol* 83: 323–326

80   Grewe M, Trefzer U, Ballhorn A, Gyufko K, Henninger H, Krutmann J (1993) Analysis of the mechanism of ultraviolet (UV) B radiation-induced prostaglandin E_2 synthesis by human epidermoid carcinoma cells. *J Invest Dermatol* 101: 528–531

81   Ashendel CL, Boutwell R, K (1979) Prostaglandin E and F levels in mouse epidermis are increased by tumor-promoting phorbol ester. *Biochem Biophys Res Commun* 90: 623–627

82   Fürstenberger G, deBravo M, Bertsch S, Marks F (1979) The effect of indomethacin on cell proliferation induced by chemical and mechanical means in mouse epidermis *in vivo*. *Res Commun Chem Pathol Pharmacol* 24: 533–541

83   Fürstenberger G, Marks F (1978) Indomethacin inhibition of cell proliferation induced by the phorbol ester TPA is reversed by prostaglandin E_2 in mouse epidermis *in vivo*. *Biochem Biophys Res Commun* 84: 1103–1111

84 Blacker KL, Williams ML, Goldyne ME (1986) 6-Ketoprostaglandin $F_{1\alpha}$ is a marker for keratinocyte-fibroblast interactions. *J Invest Dermatol* 86: 464

85 Scholz K, Fürstenberger G, Müller-Decker K, Marks F (1995) Differential expression of prostaglandin-H synthase isoenzymes in normal and activated keratinocytes *in vivo* and *in vitro*. *Biochem J* 309: 263–269

86 Müller-Decker K, Scholz K, Marks F, Fürstenberger G (1995) Differential expression of prostaglandin H synthase isozymes during multistage carcinogenesis in mouse epidermis. *Mol Carcinog* 12: 31–41

87 Maldve RE, Fischer SM (1996) Multifactor regulation of prostaglandin H synthase-2 in murine keratinocytes. *Mol Carcinog* 17: 207–216

88 Mestre JR, Subbaramaiah K, Sacks PG, Schantz SP, Tanabe T, Inoue H, Dannenberg AJ (1997) Retinoids suppress phorbol ester-mediated induction of cyclooxygenase-2. *Cancer Res* 57: 1081–1085

89 Hamberg M, Samuelsson B (1967) Oxygenation of unsaturated fatty acids by the vesicular gland of sheep. *J Biol Chem* 242: 5344–5354

90 Camp RDR, Mallet AI, Woollard PM, Brain SD, Kobza Black A, Greaves MW (1983) The identification of hydroxy fatty acids in psoriatic skin. *Prostaglandins* 26: 431–447

91 Baer AN, Costello PB, Green FA (1990) Free and esterified 13 (*R,S*)-hydroxyoctadecadienoic acids: principal oxygenase products in psoriatic skin scales. *J Lipid Res* 31: 125–130

92 Baer AN, Costello P, Green FA (1991) Stereospecificity of the products of the fatty acid oxygenases derived from psoriatic scales. *J Lipid Res* 32: 341–347

93 Buchanan MR, Haas TA, Lagarde M, Guichardant M (1985) 13-Hydroyoctadecadienoic acid is the vessel wall chemorepellant factor, LOX. *J Biol Chem* 260: 16056–16059

94 Iversen L, Fogh K, Bojesen G, Kragballe K (1991) Linoleic acid and dihomogammalinolenic acid inhibit leukotriene B_4 formation and stimulate the formation of their 15-lipoxygenase products by human neutrophils *in vitro*. Evidence of formation of anti-inflammatory compounds. *Agents Actions* 33: 286–291

95 Ku G, Thomas CE, Akeson AL, Jackson RL (1992) Induction of interleukin 1β expression from human peripheral blood monocyte-derived macrophages by 9-hydroxyoctadecadienoic acid. *J Biol Chem* 267: 14183–14188

96 Yamaja Setty BN, Berger M, Stuart MJ (1987) 13-Hydroxyoctadecadienoic acid (13-HODE) stimulates prostacyclin production by endothelial cells. *Biochem Biophys Res Comm* 146: 502–509

97 Bandyopadhyay GK, Imagawa W, Wallace D, Nandi S (1987) Linoleate metabolites enhance the *in vitro* proliferative response of mouse mammary epithelial cells to epidermal growth factor. *J Biol Chem* 262: 2750–2756

98 Glasgow WC, Eling TE (1991) Epidermal growth factor stimulates linoleic acid metabolism in BALB/c 3T3 fibroblasts. *Mol Pharmacol* 38: 503–510

99 Glasgow WC, Afshari CA, Barrett JC, Eling TE (1992) Modulation of the epidermal

growth factor mitogenic response by metabolites of linoleic and arachidonic acid in syrian hamster embryo fibroblasts. *J Biol Chem* 267: 10771–10779

100 Rao GN, Alexander RW, Runge MS (1995) Linoleic acid and its metabolites, hydroperoxy-octadecadienoic acids, stimulate c-*fos*, c-*jun* and c-*myc* mRNA expression, mitogen activated protein kinase activation and growth in rat aortic smooth muscle cells. *J Clin Invest* 96: 842–846

101 Camacho M, Godessart N, Antón R, García M, Vila L (1995) Interleukin-1 enhances the ability of cultured human umbilical vein endothelial cells to oxidize linoleic acid. *J Biol Chem* 270: 17279–17286

102 Godessart N, Camacho M, López-Belmonte J, Antón R, García M, De Moragas JM, Vila L (1996) Prostaglandin H-synthase-2 is the main enzyme involved in the biosynthesis of octadecanoids from linoleic acid in human dermal fibroblasts stimulated with IL-1β. *J Invest Dermatol* 107: 726–732

103 Loftin CD, Eling TE (1996) Prostaglandin synthase 2 expression in epidermal growth factor-dependent proliferation of mouse keratinocytes. *Arch Biochem Biophys* 330: 419–429

104 Nugteren DH, Kivits GAA (1987) Conversion of linoleic acid and arachidonic acid by skin epidermal lipoxygenases. *Biochim Biophys Acta* 921: 135–141

105 Camacho M, Vila L (1992) Biosynthesis and esterification of 13-hydroxy-eicosatetraenoic acid into phospholipids by human epidermal cells in suspensions. *J Invest Dermatol* 98: 527

106 López S, Vila L, Breviario F, de Castellarnau C (1993) Interleukin I increases 15-hydroxyeicosatetraenoic acid formation in cultured human endothelial cells. *Biochim Biophys Acta* 1170: 17–24

107 Godessart N, Vila L, Puig L, de Moragas JM (1994) Interleukin-1 increases 15-hydroxyeicosatetraenoic acid production in human dermal fibroblasts. *J Invest Dermatol* 102: 98–104

108 Morham SG, Langenbach R, Loftin CD, Tiano HF, Vouloumanos N, Jennette JC, Mahler JF, Kluckman KD, Ledford A, Lee CA et al (1995) Prostaglandin synthase 2 gene disruption causes severe renal pathology in the mouse. *Cell* 83: 473–482

109 Evans CB, Pillai S, Goldyne ME (1993) Endogenous prostaglandin E_2 modulates calcium-induced differentiation in human skin keratinocytes. *Prostaglandins Leukot Essent Fatty Acids* 49: 777–781

110 Leong J, Hughes-Fulford M, Rakhlin N, Habib A, Maclouf J, Goldyne ME (1996) Cyclooxygenases in human and mouse skin and cultured human keratinocytes: association of COX-2 expression with human keratinocyte differentiation. *Exp Cell Res* 224: 79–87

111 Hammarström S, Hamberg M, Samuelsson B, Duell AE, Stawiski M, Voorhees JJ (1975) Increased concentrations of nonesterified arachidonic acid, 12L-hydroxy-5,8,10,14-eicosatetraenoic acid, prostaglandin E_2 and prostaglandin $F_{2\alpha}$ in epidermis of psoriasis. *Proc Natl Acad Sci USA* 72: 5130–5134

112 Fogh K, Kiil J, Herlin T, Ternowitz T, Kragballe K (1987) Heterogeneous distribution of lipoxygenase products in psoriatic skin lesions. *Arch Dermatol Res* 279: 504–511

113 Woollard PM (1986) Stereochemical difference between 12-hydroxy-5,8,10,14-eicosate-traenoic acid in platelets and psoriatic lesions. *Biochem Biophys Res Commun* 136: 169–176

114 Wollard PM, Cunnigham FM, Murphy GM, Camp RD, Derm FF, Greaves MW (1989) A comparison of the proinflammatory effects of 12(R)- and 12(S)-hydroxy-5,8,10,14-eicosatetraenoic acid in human skin. *Prostaglandins* 38: 465–471

115 Woollard PM, Murphy GM, Cunningham FM, Camp RDR, Greaves MW (1988) Proin-flammatory effects of 12(R)-hydroxy-5,8,10,14-eicosatetraenoic acid in human skin. *Br J Dermatol* 118: 277

116 Bacon KB, Camp RDR (1990) Lipid lymphocyte chemoattractants in psoriasis. *Prosta-glandins* 40: 603–614

117 Otto WR, Barr RM, Dowd PM, Wright NA, Greaves MW (1989) 12-Hydroxy-5,8,10,14-eicosatetraenoic acid (12-HETE) does not stimulate proliferation of human neonatal keratinocytes. *J Invest Dermatol* 92: 683–688

118 Van de Sandt JJ, Bos TA, Rutten AA (1995) Epidermal cell proliferation and terminal differentiation in skin organ culture after topical exposure to sodium dodecyl sulphate. *In Vitro Cell Dev Biol Anim* 31: 761–766

119 Arenberger P, Kemeny L, Rupec R, Bieber T, Ruzicka T (1992) Langerhans cells of the human skin possess high-affinity 12(S)-hydroxyeicosatetraenoic acid receptors. *Eur J Immunol* 22: 2469–2472

120 Arenberger P, Kemény L, Ruzicka T (1992) Defect of epidermal 12(S)-hydroxyeicosate-traenoic acid receptors in psoriasis. *Eur J Clin Invest* 22: 235–243

121 Capdevila J, Yadagiri P, Manna S, Falck JR (1986) Absolute configuration of the hydroxyeicosatetraenoic acids (HETEs) formed during catalytic oxygenation of arachi-donic acid by microsomal cytochrome P-450. *Biochem Biophys Res Commun* 141: 1007–1011

122 Honn KV, Tang DG, Gao X, Butovich IA, Liu B, Timar J, Hagmann W (1994) 12-Lipoxygenases and 12(S)-HETE: role in cancer metastasis. *Cancer Metastasis Rev* 13: 365–396

123 Chen X-S, Kurre U, Jenkins NA, Copeland NG, Funk CD (1994) cDNA cloning, expres-sion, mutagenesis of C-terminal isoleucine, genomic structure and chromosomal local-izations of murine 12-lipoxygenases. *J Biol Chem* 269: 13979–13987

124 Takahashi Y, Ueda N, Yamamoto S (1988) Two immunologically and catalytically dis-tinct arachidonate 12-lipoxygenases of bovine platelets and leukocytes. *Arch Biochem Biophys* 266: 613–621

125 Takahashi Y, Ramesh Reddy G, Ueda N, Yamamoto S, Arase S (1993) Arachidonate 12-lipoxygenase of platelet-type in human epidermal cells. *J Biol Chem* 268: 16443–16448

126 Hussain H, Shornick LP, Shannon VR, Wilson JD, Funk CD, Pentland AP, Holtzman MJ

(1994) Epidermis contains platelet-type 12-lipoxygenase that is overexpressed in germinal layer keratinocytes in psoriasis. *Am J Physiol* 266: C243–C253

127 Krieg P, Kinzig A, Ress-Löschke M, Vogel S, Vanlandingham B, Stephan M, Lehmann WD, Marks F, Fürstenberger G (1995) 12-Lipoxygenase isoenzymes in mouse skin tumor development. *Mol Carcinog* 14: 118–129

128 Van Dijk KW, Steketee K, Havekes L, Frants R, Hofker M (1995) Genomic and cDNA cloning of a novel mouse lipoxygenase gene. *Biochim Biophys Acta* 1259: 4–8

129 Funk CD, Keeney DS, Oliw EH, Boeglin WE, Brash AR (1996) Functional expression and cellular localization of a mouse epidermal lipoxygenase. *J Biol Chem* 271: 23338–23344

130 Kinzig A, Fürstenberger G, Bürger F, Vogel S, Müller-Decker K, Mincheva A, Lichter P, Marks F, Krieg P (1997) Murine epidermal lipoxygenase (Aloxe) encodes a 12-lipoxygenase isoform. *FEBS Lett* 402: 162–166

131 Antón R, Abián J, Vila L (1995) Characterization of arachidonic acid metabolites through the 12-lipoxygenase pathway in human epidermis by high-performance liquid chromatography and gas chromatography/mass spectrometry. *J Mass Spectrom and Rapid Commun Mass Spectrom* S169–S182

132 Fruteau de Laclos B, Borgeat P (1988) Conditions for the formation of the oxo derivatives of arachidonic acid from platelet 12-lipoxygenase and soybean 15-lipoxygenase. *Biochim Biophys Acta* 958: 424–433

133 Pace-Asciak CR, Granström E, Samuelsson B (1983) Arachidonic acid epoxides. Isolation and structure of two hydroxy epoxide intermediates in the formation of 8, 11, 12- and 10,11,12-trihydroxyeicosatrienoic acids. *J Biol Chem* 258: 6835–6840.

134 Pace-Asciak CR (1984) Hemoglobin- and hemin-catalyzed transformation of 12L-hydroperoxy-5,8,10,14-eicosatetraenoic acid. *Biochim Biophys Acta* 793: 485–488

135 Reynaud D, Demin P, Pace-Asciak CR (1994) Hepoxilin A_3 formation in the rat pineal gland selectively utilizes (12S)-hydroperoxy-eicosatetraenoic acid (HPETE), but not (12R)-HPETE. *J Biol Chem* 269: 23976–23980

136 Vasiljeva LL, Manukina TA, Demin PM, Lapitskaja MA, Pivnitsky KK (1993) Synthesis, properties and identification of epimeric hepoxilins (–)-(10R)-B_3 and (+)-(10S)-B_3. *Tetrahedron* 49: 4099–4106

137 Demin PM, Pivnitsky KK, Vasiljeva LL, Pace-Asciak CR (1994) Synthesis of methyl [5,6,8,9,14,15-$^3H_6$]-hepoxilin B_3 and its conversion into methyl [5,6,8,9,14,15-$^3H_6$]-hepoxilin A_3. *J Label Comp Radiopharm* 34: 221–230

138 Laneuville O, Corey EJ, Couture R, Pace-Asciak CR (1991) Hepoxilin A_3 increases vascular permeability in the rat skin. *Eicosanoids* 4: 95–97

139 Wang M-M, Demin PM, Pace-Asciak CR (1996) Epimer-specific actions of hepoxilins A_3 and B_3 on PAF- and bradykinin-evoked vascular permeability in the rat skin *in vivo*. *Adv Exp Med Biol* 416: 239–241

140 Van Wauwe J, Coene M-C, Van Nyen G, Cools W, Goossens J, Le Jeune L, Lauwers W, Janssen PAJ (1991) NADPH-dependent formation of 15- and 12-hydroxy-eicosatrienoic acid from arachidonic acid by rat epidermal microsomes. *Eicosanoids* 4: 155–163

141 Murphy RC, Falf JR, Lumin S, Yadagiri P, Zirrolli JA, Balazy M, Masferrer JL, Abraham NG, Schwartzman ML (1988) 12(R)-hydroxyeicosatrienoic acid: a vasodilator cytochrome P450-dependent arachidonate metabolite from the bovine corneal epithelium. *J Biol Chem* 263: 17197–17202

142 Masferrer J, Murphy RC, Pagano PJ, Dunn MW, Schwartzman ML (1989) Ocular effects of a novel cytochrome P450-dependent arachidonic acid metabolite. *Invest Ophthalmol Vis Sci* 30: 454–460

143 Conners MS, Schwartzman ML, Quan X, Heilman E, Chauhan K, Falck JR, Godfrey HP (1995) Enhancement of delayed hypersensitivity inflammatory reactions in guinea pig skin by 12(R)-hydroxy-5,8,14-eicosatrienoic acid. *J Invest Dermatol* 104: 47–51

144 Pace-Asciak CR, Lee WS (1989) Purification of hepoxilin epoxide hydrolase from rat liver. *J Biol Chem* 264: 9310–9313.

145 Pace-Asciak CR, Laneuville O, Chang M, Reddy CC, Su WG, Corey EJ (1989) New products in the hepoxilin pathway: isolation of 11-glutathionyl hepoxilin A_3 through reaction of hepoxilin A_3 with glutathione S-transferase. *Biochem Biophys Res Commun* 163: 1230–1234

146 Pace-Asciak CR, Laneuville O, Su WG, Corey EJ, Gurevich N, Wu P, Carlen PL (1990). A glutathione conjugate of hepoxilin A_3: formation and action in the rat central nervous system. *Proc Natl Acad Sci USA* 87: 3037–3041

147 Antón R, Puig L, Esgleyes T, de Moragas JM, Vila L (1998) Occurrence of hepoxilins and trioxilins in psoriatic lesions. *J Invest Dermatol* 110: 303–350

148 Capdevila JH, Karara A, Waxman DJ, Martin MV, Falck JR, Guengerich FP (1990) Cytochrome P-450 enzyme-specific control of the regio- and enantiofacial selectivity of the microsomal arachidonic acid epoxygenase. *J Biol Chem* 265: 10865–10871

149 Garssen G J, Veldink GA, Vliegenthart JFG, Boldingh J (1976) The formation of *threo*-11-hydroxy-*trans*-12:13-epoxy-9-*cis*-octadecenoic acid by enzymic isomerisation of 13-L-hydroperoxy-9-*cis*,11-*trans*-octadecadienoic acid by soybean lipoxygenase-1. *Eur J Biochem* 62: 33–36

150 Carroll MA, Schwartzman M, Sacerdoti D, McGiff JC (1988) Novel renal arachidonate metabolites. *Am J Med Sci* 295: 268–274

151 Rubbo H, Radi R, Trujillo M, Telleri R, Kalyanaraman B, Barnes S, Kirk M, Freeman BA (1994) Nitric oxide regulation of superoxide and peroxynitrite-dependent lipid peroxidation. *J Biol Chem* 269: 26066–26075

152 Laskey RE, Mathews WR (1996) Nitric oxide inhibits peroxynitrite-induced production of hydroxyeicosatetraenoic acids and F_2-isoprostanes in phosphatidylcholine liposomes. *Arch Biochem Biophys* 330: 192–198

153 Kolb-Bachofen V, Fehsel K, Michel G, Ruzicka T (1994) Epidermal keratinocyte expression of inducible nitric oxide synthase in skin lesions of psoriasis vulgaris. *Lancet* 344: 139–140

154 Sirsjö A, Karlsson M, Gidlöf A, Rollman O, Törmä H (1996) Increased expression of inducible nitric oxide synthase in psoriatic skin and cytokine-stimulated cultured keratinocytes. *Br J Dermatol* 134: 643–648

155 Nigam S, Müller S, Pace-Asciak CR (1993) Hepoxilins activate phospholipase D in the human neutrophil. *Dev Oncol* 71, 249–252

156 Reynaud D, Demin P, Pace-Asciak CR (1996) Hepoxilin A$_3$-specific binding in human neutrophils. *Biochem J* 313: 537–541

157 Dho S, Grinstein S, Corey EJ, Su WG, Pace-Asciak CR (1990) Hepoxilin A$_3$ induces changes in cytosolic calcium, intracellular pH and membrane potential in human neutrophils. *Biochem J* 266: 63–68

158 Laneuville O, Reynaud D, Grinstein S, Nigam S, Pace-Asciak CR (1993) Hepoxilin A$_3$ inhibits the rise in free intracellular calcium evoked by formyl-methionyl-leucyl-phenylalanine, platelet-activating factor and leukotriene B$_4$. *Biochem J* 295: 393–397

159 Tuschil A, Lam C, Haslberger A, Lindley I (1992) Interleukin-8 stimulates calcium transients and promotes epidermal cell proliferation. *J Invest Dermatol* 99: 294–298

160 Pace-Asciak CR, Martin JM (1984) Hepoxilin, a new family of insulin secretagogues formed by intact rat pancreatic islets. *Prostaglandins Leukot Med* 16: 173–180

161 Pace-Asciak CR, Martin JM, Corey EJ (1986) Hepoxilins potential endogenous mediators of insulin release. *Prog Lipid Res* 25: 625–628

162 Piomelli D, Shapiro E, Feinmark SJ, Schwartz JH (1987) Metabolites of arachidonic acid in the nervous system of Aplysia: possible mediators of synaptic modulation. *J Neurosci* 7: 3675–3686

163 Pace-Asciak CR (1994) Hepoxilins: a review on their cellular actions. *Biochim Biophys Acta* 1215: 1–8

Strategies for the analysis of fatty acid mediators of inflammation

Anthony I. Mallet

St. John's Institute of Dermatology, UMDS, University of London, St. Thomas' Hospital, Lambeth Palace Road, London SE1 7EH, England

Introduction

The role of fatty acids in mediating inflammatory processes in human skin is clearly established. The mechanisms by which they operate extend from gross physico-chemical changes, e.g. in cell membrane properties, to highly specific, receptor-mediated, cell activation. The population of fatty acid molecules is in a state of flux, free acids being continuously taken up into specific lipid pools and released from these. In addition, they are subject to metabolic and oxidative modifications, both in the free and bound forms, leading to a wide variety of molecular species, several of which have important biological activities. A number of excellent practical texts for use on the laboratory bench exist, in which the methods for fatty acid extraction, purification and analysis are described in detail, including the books by Christie and Hamilton and Hamilton [1,2]. In this review I have taken examples of protocols from a wide range of human matrices, as the methods describe therein apply equally to samples of dermal relevance.

Molecular sources of fatty acids

Most of the common fatty acids in animal systems are derived from long chains with even-numbered carbon atoms, but fatty acids with an uneven number of carbon atoms do occur, for example in fatty acids of bacterial origin. From these "saturated" fatty acids oxidative enzyme systems produce a variety of unsaturated acids with one or more double bonds, most usually separated by a single methylene unit. The geometrical configuration of these bonds is almost universally *cis* (Z). Figure 1 illustrates some of the fatty acids most frequently studied in connection with inflammation in skin; this also defines the nomenclature for these molecules and the numbering systems most commonly used. Many of the fatty acid molecules of interest in studies of inflammation are derived from two principal pathways starting with

Fatty Acids and Inflammatory Skin Diseases, edited by J.-M. Schröder
© 1999 Birkhäuser Verlag Basel/Switzerland

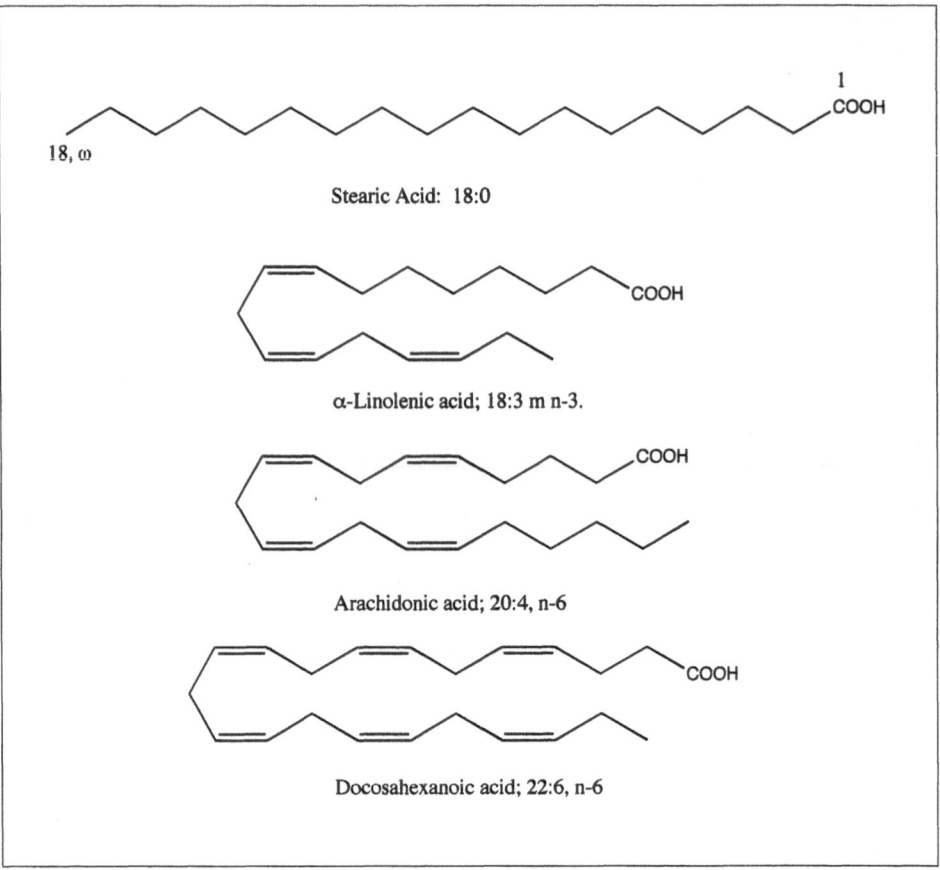

Figure 1
Five common fatty acids from which mediators of inflammation are derived. The numbering system refers to the total carbon chain length, the number of double bonds and the number of carbon atoms from the terminal carbon (ω) at which the unsaturation begins (n3 or n6).

linoleic and α-linolenic acids (Fig. 2). From these two pathways the majority of molecules described in the literature on dermal inflammation are derived.

Metabolism of these fatty acids, especially the unsaturated acids such as arachidonic acid (AA), by oxidation with appropriate enzymes or by free radical attack, leads to the formation of a variety of substituted, modified and biologically active molecules. The most common substitutents found are hydroxyl, oxo and carboxylic acid groups but aldehydes, epoxy groups, and chain shortening and lengthening are also seen. Figure 3 illustrates some common oxidised fatty acid derivatives mostly,

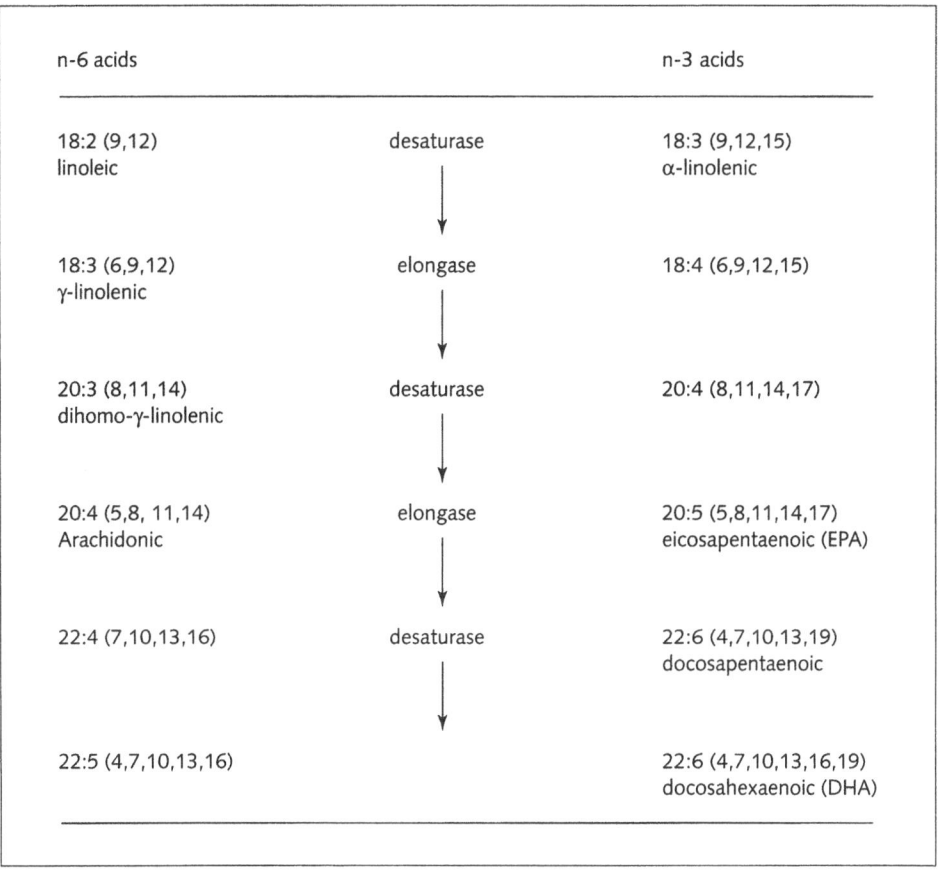

Figure 2
The two principal pathways for unsaturated fatty acid synthesis.

but not exclusively, derived from AA, which are described in studies into the molecular mechanisms of inflammatory disorders.

While fatty acids do exist in biological matrices in the free form, a large proportion will be found esterified to glycerol, phosphoglycerides and other lipids such as cholesterol (Fig. 4). These lipids are the principal structural units of all cell membranes, and the nature and shapes of the fatty acid constitutents are the major determinants of important membrane properties such as its fluidity. The release of fatty acids from esterified lipid pools and their incorporation into these forms is tightly controlled by lipases (phospholipases) and transesterifying enzymes.

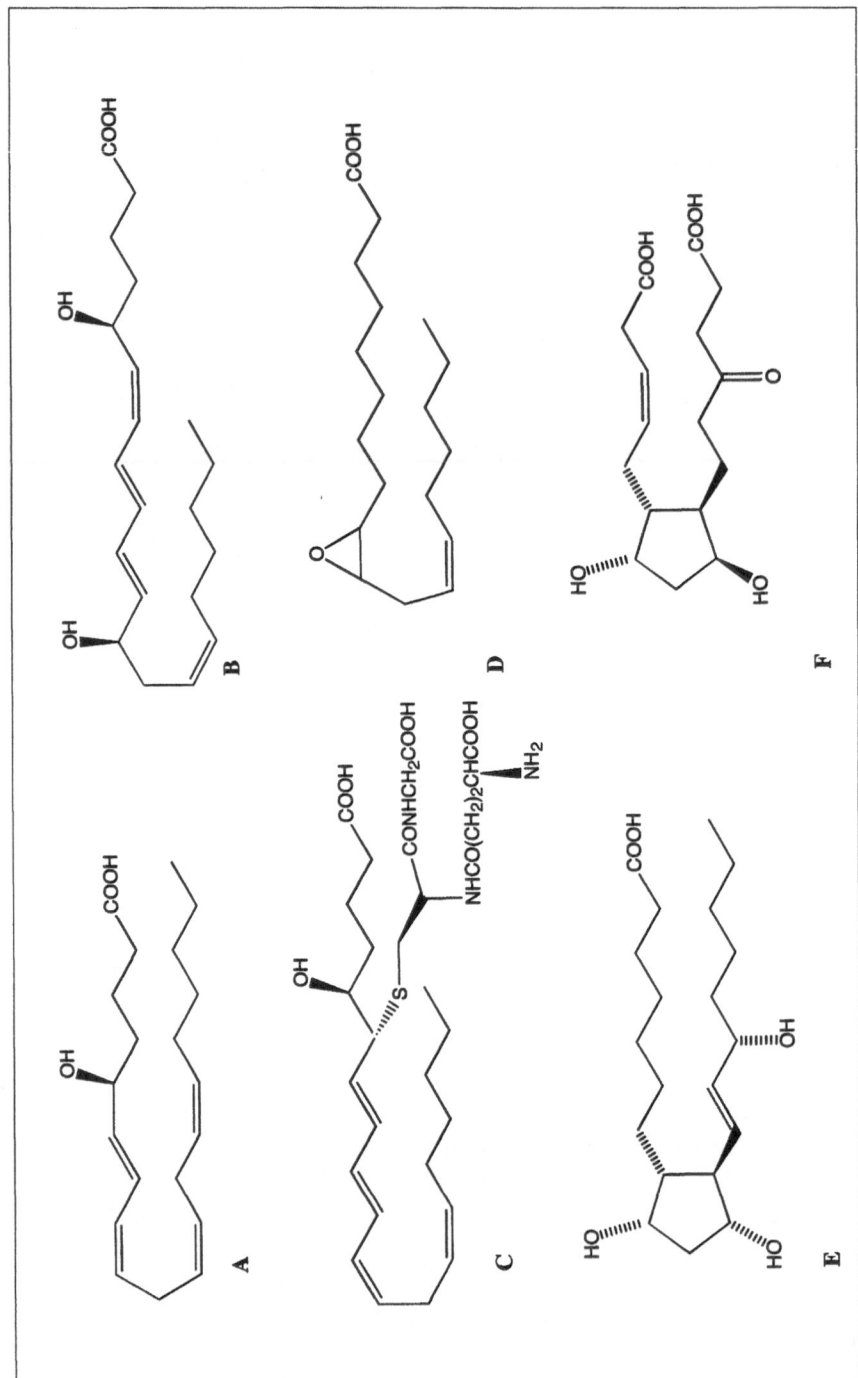

Figure 3
Oxidised fatty acid metabolites with important biological activity; A, 5-HETE; B, LTB₄; C, LTC₄; D, leukotoxin; E, PGF₂ₐ; and F, PGD-M.

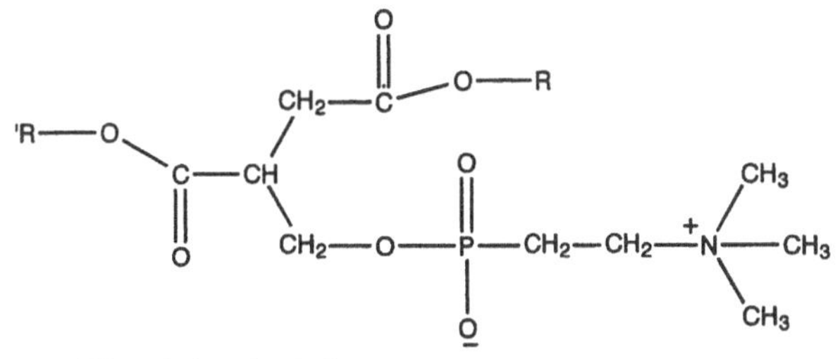

Di-acyl-phosphocholine

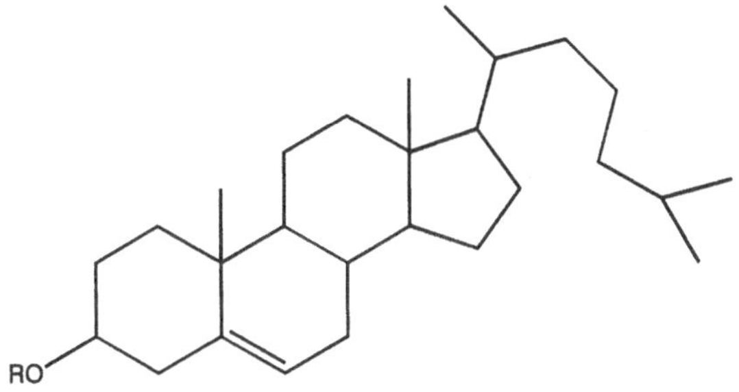

Cholesterol ester

Figure 4
Acids are commonly found in membranes esterified to phospholipids and cholesterol.

Extraction of free fatty acids

All unsaturated fatty acids are susceptible to adventitious oxidation introduced during the process of extraction, purification and storage. The addition of antioxidants to the extraction media is commonly advised, and butylated hydroxytoluene (BHT) is the most commonly recommended reagent. The addition of enzyme inhibitors is also frequently recommended; the addition of a cyclooxygenase inhibitor such as indomethacin can be of great value in suppressing unwanted prostaglandin formation during the workup of a biological matrix. Solvents to be employed must be peroxide-free and should also be examined to ensure that they are themselves free of lipid contaminants. Removal of solvents and derivatising agents is frequently performed by the use of a gentle stream of nitrogen. This can lead to losses of sample

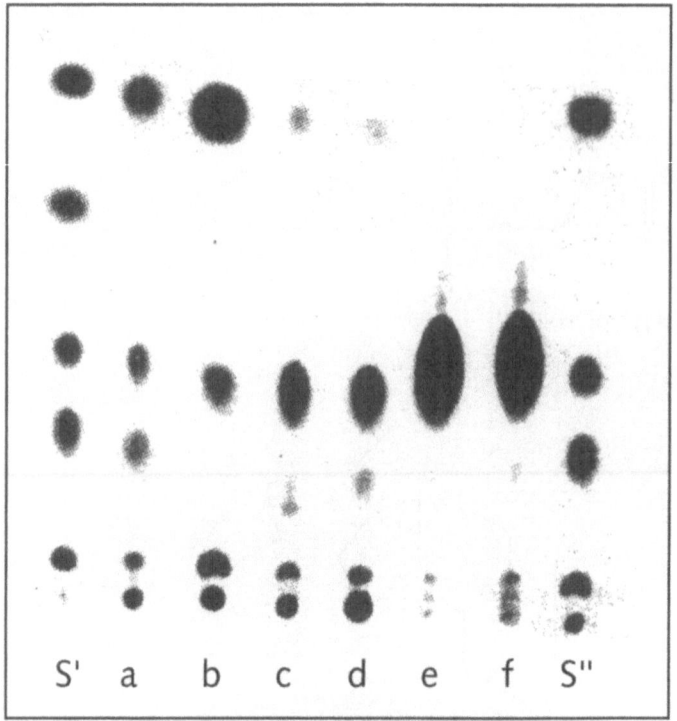

Figure 5
TLC of human tissue lipids. Silica gel developed with heptane:diethylether:acetic acid. Lane S' standard cholesterol, oleic acid, triolein, oleic acid methyl ester and cholesterol oleate; lane S" also with phosphatidyl choline mixture. Lanes a–f, various tissue extracts. Developed by charring with sulphuric acid.

by splashing and volatilisation, and the use of a vacuum centrifugation apparatus is preferable. Dried-down samples should immediately be redissolved in an appropriate solvent and not left dry for longer than necessary. All samples should be stored in the dark, preferably at below −70° C, with precautions to exclude air by the use of a nitrogen or, better, argon purge.

If all the fatty acids are to be analysed, prior separation of the different lipid classes is not necessary, and a simple hydrolysis using alcoholic potassium hydroxide followed by acidification will provide a medium containing all the fatty acids in a free form [3]. If performing a specific analysis of each esterified form of the fatty acid is required, a lipid class separation will need to be carried out. Several procedures have been described for this, including high-performance liquid chromatography (HPLC), solid-phase extraction (SPE) cartridges and thin layer chromatography

(TLC). The method described by Kaluzny [4] is especially convenient and elutes all lipid classes separately using aminopropyl SPE columns. For a quick survey of the classes present, TLC is very convenient (Fig. 5) and is used for purifying a limited number of samples. Lipids can be detected by iodination, charring with acids, fluorescent developing agents or by radioactivity where an appropriate substrate has been processed. It is frequently difficult to recover the analyte from the TLC plate free of silica, and further solvent extraction is necessary. TLC methods for lipids have been well reviewed by Henderson and Tocher [5].

Whichever source of the fatty acids is to be treated, the extraction procedures all employ a mixture of immiscible aqueous and nonpolar solvents in such proportions as to extract all lipids into the organic rich layer and leave all more polar constitutents in the aqueous medium. The classical protocols are those described by Folch, Lees and Stanley [6], and by Bligh and Dyer [7]. The principle behind these methods rests on the formation of a two-phase system, when appropriate volumes of chloroform, methanol and water are mixed. In reproducing the protocols described in the literature, it is essential to conform precisely to the volumes described; in the system proposed by Bligh and Dyer, the final overall solvent composition should be chloroform:methanol:water in the proportions 2:2:1.8 v/v/v at 20°C. Alternative solvent systems include the use of 2-propanol:hexane:water mixtures which have the advantages of neutralising any enzyme systems efficiently and of avoiding the problems of toxic and reactive side products in chloroform. The above initial extraction can be performed on plasma, skin blister exudates, urine, cell culture medium and macerated tissue. Prior precipitation of protein in biological fluids is often helpful, as this prevents blockages in later steps and the incomplete separation of liquid layers in two-phase extraction systems. However, if the matrix is relatively free of excess lipid and only free fatty acids are required (e.g. cell culture supernatant), direct extraction at an acid pH can be performed with a nonpolar hydrocarbon solvent.

The hydrolysis of bound fatty acids can be performed by both acid and alkaline hydrolysis, by transesterification reactions or by the use of specific enzyme preparations such as phospholipase A_2. Purification of the lipid extract to isolate the fatty acids is carried out by acidification to a pH below the pK_a of the target analytes – usually below pH 3.5 will suffice – followed by extraction from an aqueous medium into the least polar solvent that will extract the fatty acids involved. If the only fatty acids are nonpolar, a solvent such as heptane is recommended. If more polar molecules have to be extracted, it may be necessary to use mixtures of hydrocarbon solvents with polar modifiers such as chloroform, methylene chloride, ethyl acetate or 2-propanol. Alternatively, it is becoming more common to perform a selective extraction by the use of SPE cartridges packed with reversed-phase (RP) materials. These have the advantage of providing rapid, clean extractions which can be automated for multiple sample analyses. As an illustration of the above, Figure 6 below outlines a procedure used routinely in our laboratories for the extraction of

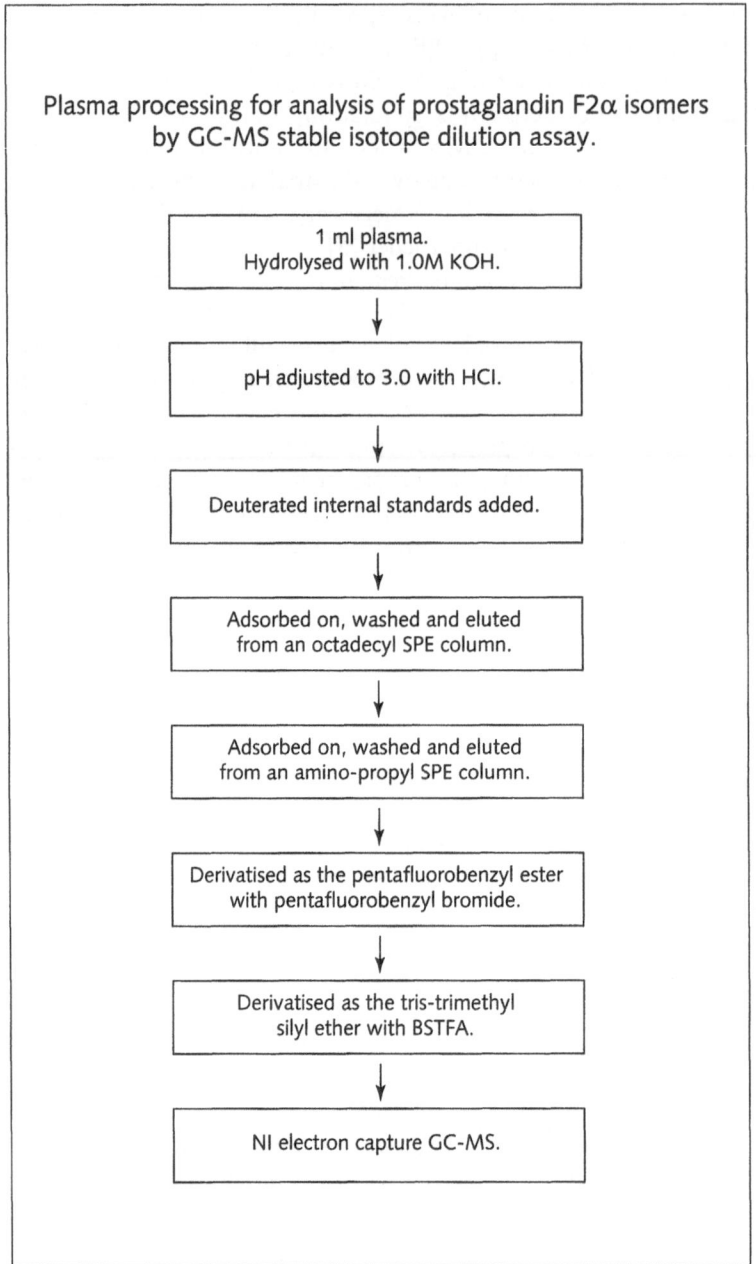

Figure 6
Flow chart for the extraction and purification of prostaglandin $F_{2\alpha}$ isomers from plasma samples.

prostaglandin-like molecules from a variety of matrices. The literature offers examples of nonpolar acid extraction [8], hydroxy fatty acid extractions [9] and hydroperoxide fatty acid extractions [10, 11].

Separation and derivatisation of fatty acids

Table 1 shows some common procedures employed for this reaction. The formation of the methyl ester is preferred over longer-chain alkyl homologues unless other considerations are important, for example electron-capturing properties of fluorinated aryl groups. Direct derivatisation by transesterification using BF_3–CH_3OH can also be used on the parent esterified lipids. Hydroxyl substitution on the fatty acid chain needs protection for gas chromatography (GC) separation, and this is most commonly performed by alkylation using a silylation reagent leading to the formation of a silyl ether derivative (Fig. 7). Variation in the nature of the alkyl groups attached to the silicon atom provides useful control of subsequent mass spectrometric properties [9].

In certain cases the presence of double bonds in the chain can lead to instablity in the molecule, and several authors have described small-scale hydrogenation methods to saturate the fatty acid chain. This procedure is frequently employed in work on AA metabolites and is illustrated by the work of Murphy et al. [19] and Woollard and Mallet [20].

Table 1 - Fatty acid esterification reagents suitable for GC analysis

Reaction	Reagent	Refs
Alkyl esters		
R–COOH → R–COOCH$_3$	Diazo methane: CH_2N_2	[12]
	CH_3OH/HCl (CH_3COCl, SO_2Cl_2)	[13]
	BF_3/CH_3OH or CH_3OH/KOH	[14]
	(efficient in transesterification of e.g. glycerides)	
	Chloroformates	[15]
R–COOH → R–COOR'	R'OH/HCl	[16]
	Alkyl ammonium hydroxide	
Aryl esters		
R–COOH → R–COO $CH_2C_6F_5$	$C_6F_5CH_2Br/Base/CH_3CN$	[17]
Silyl esters		
R–COOH → R–COO–Si(R')	BSTFA or MTBSTFA	[18]
R'= $(CH_3)_3$ or $(CH_3)_2.C_4H_9$		

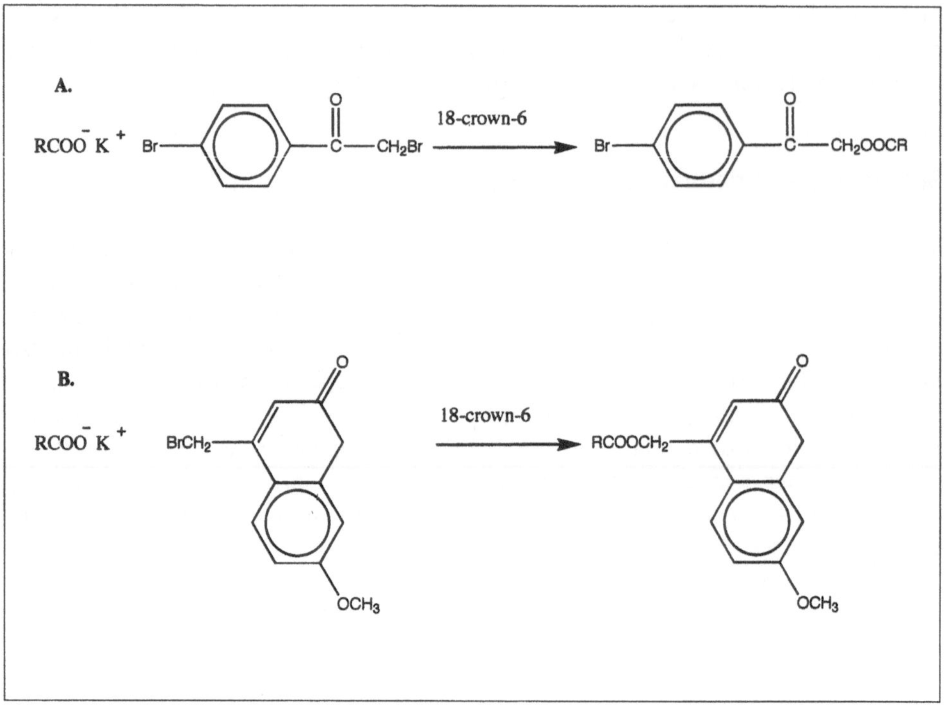

Figure 7
Derivatisation reagents for HLPC detection. (A) Bromoacetophenone ester; UV detection λ_{max} = 254 nm [23]. (B) Coumarin derivative, fluorophore excitation λ 323 nm, emission λ 395 nm [24].

Keto groups, on account of their propensity to partially form enolates and hence mixtures when the keto acid is treated with silylation reagents, can be "fixed" by the formation of the corresponding oxime. Two geometric isomers are formed in this process, and this sometimes leads to problems with the GC separation. However, the formation of oximes can be carried out in aqueous media, and several authors have described combined derivatisation and specific extraction procedures for ketonic species. Table 2 summarises some of the more common reactions employed for the formation of hydroxyl and oxo derivatives. A number of texts and reviews of derivatisation procedures exist [18, 21].

In the case of HPLC detection, if no appropriate and strongly ultraviolet (UV)-absorbing group is present, it is possible to introduce a UV chromophore or a fluorescent derivative group by esterification of the carboxyl group. Two common reagents are shown in Figure 6 [23, 24].

Table 2 - Derivatisation reactions for hydroxyl and oxo groups

Reaction	Reagent	Refs
Hydroxyl acylation		
R–OH → R–OCOCH$_3$	Acyl anhydrides/ chlorides + base	[18]
	Carboxylic acid + trifluoroacetic anhydride	[22]
Hydroxyl silyl ethers:		
R–OH → R–O–Si(R')$_3$		
R'= (CH$_3$)$_3$ or (CH$_3$)$_2$.C$_4$H$_9$	BSTFA or MTBSTFA	[18]
Ketone oximation:		
(R)$_2$C=O → (R)$_2$C=N–OR'	R'O–NH$_2$/base	[18]

Separation of fatty acids

GC is the longest-established method for fatty acid separation. The most common detector in this work is the flame ionisation detector, in which the analyte is combusted, and hence the only means of sample identification is by comparison with known standards of chromatographic retention times. This difficulty can be surmounted by the use of mass spectrometry (MS) as a detector (see below). For the analysis of a mixture of fatty acid methyl esters, MS is the most efficient method available, providing speed of analysis with efficiency of separation and sensitivity. Several authors have reviewed the use of GC for lipid analysis [8, 25, 26]. The analysis of skin surface fatty acids has been described by Patel and Noble [27].

Modern GC columns are able to perform at elevated temperatures, in excess of 360° C, and can be obtained with stationary phases which are able to offer specific fatty acid separations, such as the difficult task of resolving the various unsaturated C18 components. Figure 8 shows a survey of the normal free fatty acids present in human plasma.

Prostaglandin, leukotriene and other hydroxylated metabolites of AA can all be separated after derivatisation by appropriate GC procedures, though several authors have reported problems with thermal decomposition unless the unsaturation is removed by hydrogenation. GC cannot be employed when the molecules are too big and/or polar for efficient volatilisation. Typically, the conjugated catabolites such as glucuronides and sulphates and the LTC, D and E series of sulpho-peptide leukotrienes are not amenable to this method. In such cases the use of HPLC is to be preferred [19, 28]

HPLC, reviewed for lipid analysis by Sewell [29], has a number of advantages over GC in that it can be carried out over a wide scale of sample size (from prepar-

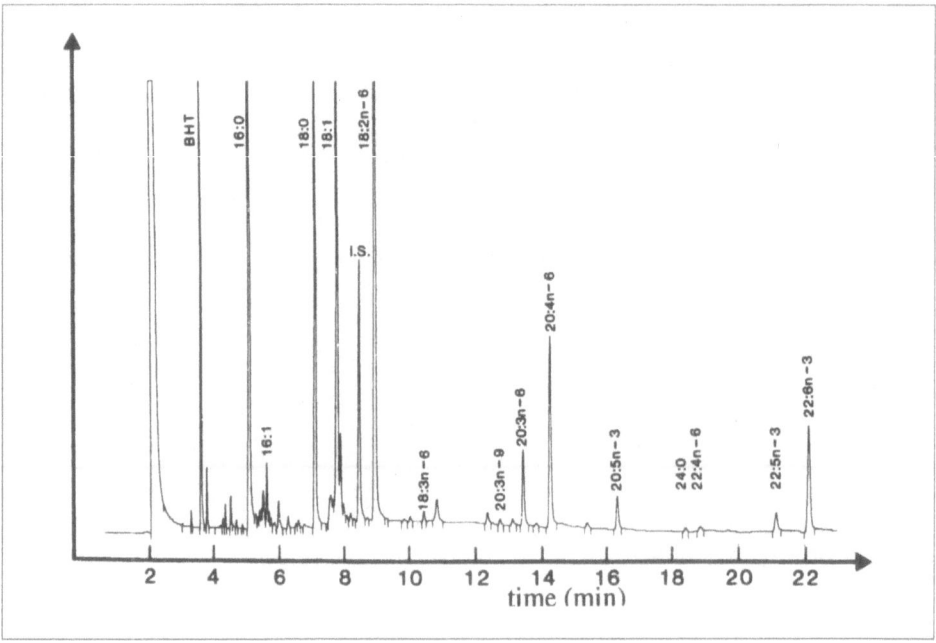

Figure 8
The chromatogram shows the separation that can be achieved by capillary GC for a mixture of fatty acids, derivatised as their methyl esters, extracted from a normal plasma sample.

ative at the milligram scale down to microcapillary where subfemtogram samples can be analysed). Provided the analyte can be detected, no derivatisation is needed, and modern HPLC systems, coupled with MS are removing the need for chemical modification of the samples. Unlike GC systems, virtually the entire sample can be applied to the column and, if a nondestructive detector is used, can be recovered after separation. The principal drawback lies in the inherently lower separation efficiencies of HPLC compared with GC, but newer techniques such as capillary electrophoresis [30] are making significant improvements in this area.

There are three major mechanisms of HPLC separation used in lipid analysis: straight-phase, using unmodified silica packed columns with solvents of increasing polarity; RP-HPLC, with packings of silica whose surface has been modified with nonpolar organic molecules and using aqueous-based solvents with increasing organic content; and ion-exchange packings, where variation of ionic strength and pH can effect selective elution from the column. It is almost always necessary to use a gradient of solvent composition in HPLC to maintain good peak shape and effect separation in reasonably short times.

HPLC has proved of great use in lipid class separation but appears to be less useful in analysing individual unsaturated fatty acids in complex mixtures. Within a given class of phospholipids, for example, the individual fatty acid-containing species can be separated by careful RP-HPLC, and the review by Olsson and Salem [31] provides good examples of this. When combined with MS (see HPLC-MS below), individual fatty acid components which are not completely separated can be distinguished by the mass spectrometer's response. Figure 9 shows the results of an analysis of a set of standard phospholipids separated by RP-HPLC with MS detection [32].

HPLC has played an important role in the analysis of arachidonic and related unsaturated fatty acid metabolites and in the identification of biologically active compounds as, for example, in the work of Borgeat et al. [28].

Detection of fatty acids and quantitation

Qualitative analysis

The use of spectrophotometric methods to detect substances eluting from HPLC columns has already been discussed. At a single fixed wavelength the classical UV detector can be uninformative as to the structure of the eluting species. The use of several wavelengths simultaneously or, better, the array diode detector in which a full scanned spectrum can be obtained can lead to much greater levels of information. Schwenk and Schroder [33] in the structural determination of an oxo-AA metabolite have made use of off-line UV spectra combined with λmax shifts in acid and base to show the presence of the α,β-unsaturated ketone entity in the molecule. Experiments like these are rapidly performed and provide useful structural information without any loss of sample. The hydroxyeicosatetraenoic acids (HETEs) all show the characteristic diene absorption at around 240 nm, and the lipoxins, trihydroxy AA metabolites with a conjugated tetraene system, show strong absorbance at 302 nm in the UV. Chavis et al. [34] have shown the use of RP-HPLC in the separation of these molecules, and Wheelan et al. have investigated the metabolism of LTB_4 by keratinocytes using similar techniques [35]. A recent study by Reynaud and Pace-Asciak has employed elegant HPLC separation methods for an investigation of the hepoxilins [36].

In the case of GC the detector is, again, structurally insensitive if a flame ionisation or electron capture detector is employed. In order to maximise the information content of these analyses, it is becoming increasingly common to use mass spectrometry linked on-line to the separation process. GC-MS has been employed for well over two decades for the identification and quantitation of volatile substances. The principle of the technique lies in the ionisation of the molecules in a vacuum chamber. The ionisation process is an energetic one, and in addition to the molecule

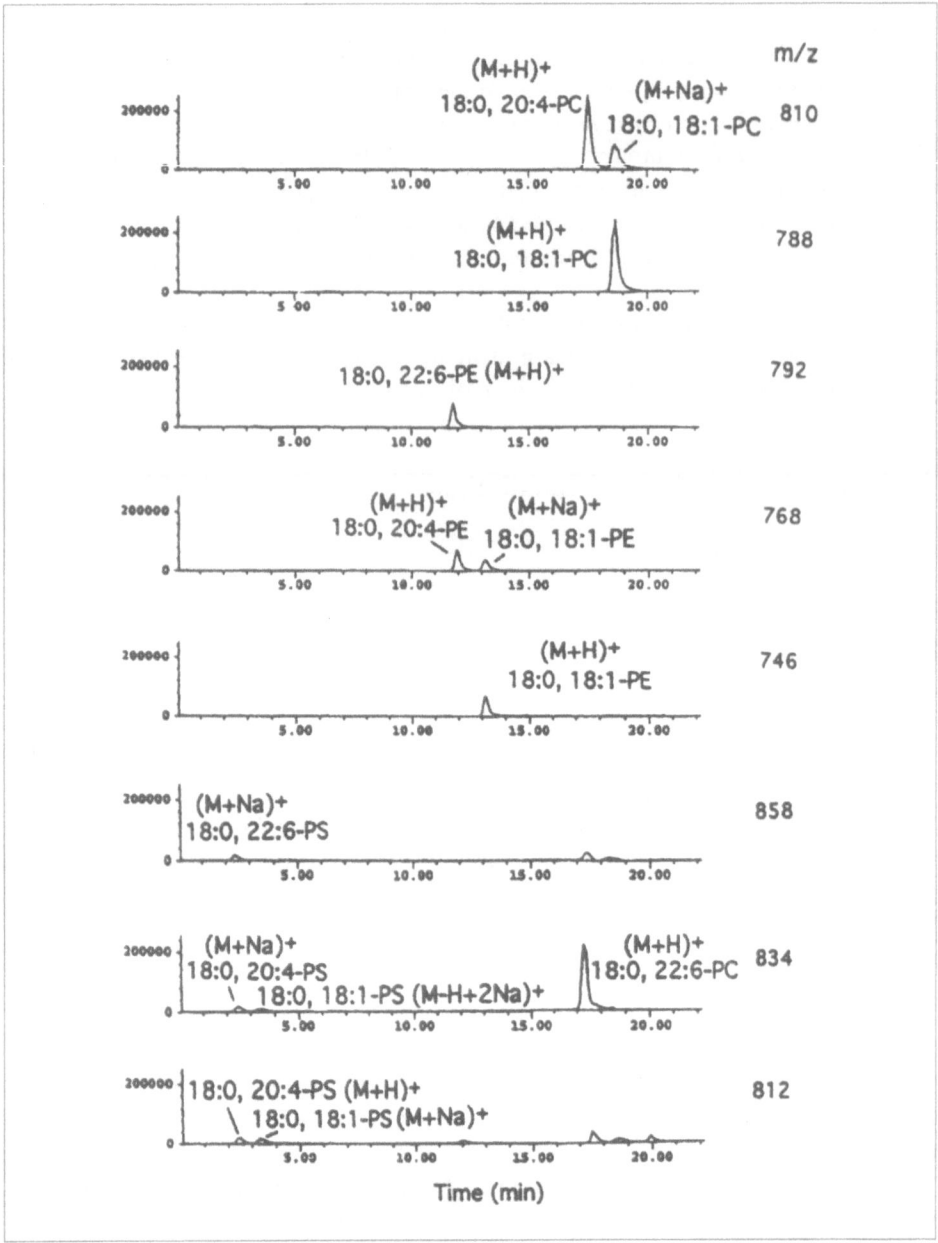

Figure 9

RP-HPLC-MS separation of individual phosphocholines (PC), phosphoethanolamine (PE) and phosphoserine (PS) species. Individual fatty acid components can be observed. Reproduced from [32] with permission.

itself several fragments may also be present. The ion is sent through electric and or magnetic fields in a way that causes the ions of different mass to separate. These are then detected, and parameters for the masses and abundances of the various ions are obtained. A plot of these is a so-called mass spectrum. The combination of analyte detection by elution time from a chromatographic system and also by its mass provides for very high levels of specificity, while the efficiency of modern ion detection systems provides high sensitivities.

The pattern of fragment ions reflects the weakest bonds in the original molecule, and hence is often structurally informative and of great use in the determination of structures of novel compounds. Nearly all the AA metabolites from the prostaglandins onwards have been characterised, at least in the initial stages, by mass spectrometry. When mass spectrometry is combined with a judicious use of simple derivatisation, a great deal of information can be obtained, and good accounts using combinations of these techniques can be found in publications by Murphy [37, 38], Evershed [39] and Christie [40]. The analysis of the positions of the unsaturated bonds in fatty acids is vital, and Harvey [41] and others [42] have published derivatisation and mass spectrometric methods for this form of analysis.

In 1995 Anton et al. [43] published an elegant study of the products of epidermal metabolism of exogenous arachidonate. A combination of straight-phase and RP-HPLC and GC-MS after appropriate derivatisation led to the characterisation of seven products of the 12-lipoxygenase pathway. In our laboratories we have made use of GC-MS in the analysis of lipoxins [44], and Figure 10 shows the mass spectrum of LXA4 and some of the structurally characteristic fragmentations. Final proof of structure has to be obtained by nuclear magnetic resonance (NMR) and synthesis, especially for definition of the absolute stereochemistry of the molecule, but milli- or micromolar quantities of pure material are required for this.

In recent years it has become equally common, and simple, to link HPLC eluants to a mass spectrometer, and instrumentation for this is now widely available. There are limits to the flow rates that can be accepted by the mass spectrometer and to the ionisation processes that can be effected. The latter means that more polar and easily ionised species are favoured, but these are just those that cannot be accepted by GC-MS systems, and the two techniques are complementary. The information content of an LC-MS experiment can be considerably enhanced by the use of tandem mass spectrometry (MS-MS), described by Murphy [37], and Nakamura et al. have published an elegant study of the use of these techniques in a study of the epoxy- and hydroxyeicosatetraenoic acids esterified into phospholipids [45] (Fig. 11). Determination of the structure of novel substances with defined biological activity is greatly assisted by on-line chromatography combined with UV and MS detection as well as simultaneous bioassay, and a recent paper by Watson et al. brings together these techniques to determine the structures of biologically active oxidised fatty acids esterified to phospholipids [46] and to confirm the molecular identity of biologically active entities extracted from an *in vivo* source.

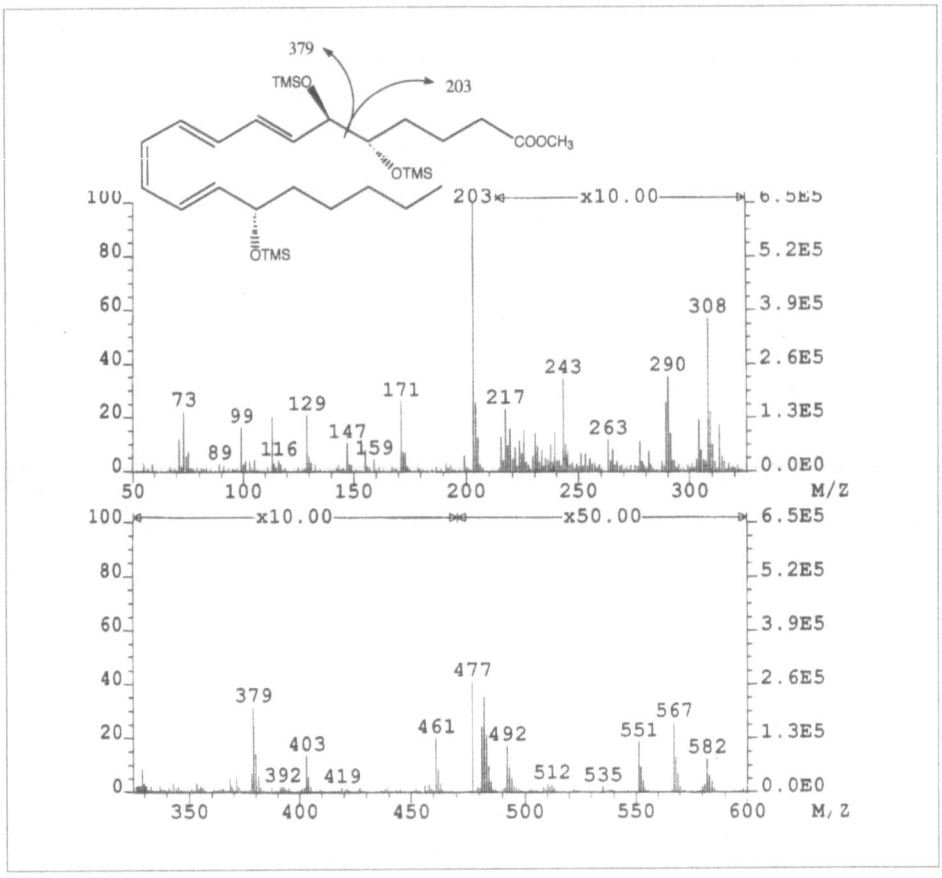

Figure 10
EI mass spectrum and structure of the lipoxin A4 molecule. The methyl ester tri-TMS ether derivative has a molecular weight of 582. Characteristic fragment ions at masses 203 and 379 can be observed.

Quantitative analysis

The ability conferred by a mass spectrometer to differentiate between molecules of differing mass has led to the widespread use of "stable isotope dilution" techniques for the quantitation of target analytes [47–50]. In this technique use is made of a stable isotope-substituted analogue of the analyte molecule, usually with ^{2}H, ^{13}C or ^{18}O atoms, added at the beginning of the extraction and purification process to compensate for any losses that may occur. In the GC-MS or LC-MS system the ratio of the mass of the analyte and that of the internal standard are determined, and the

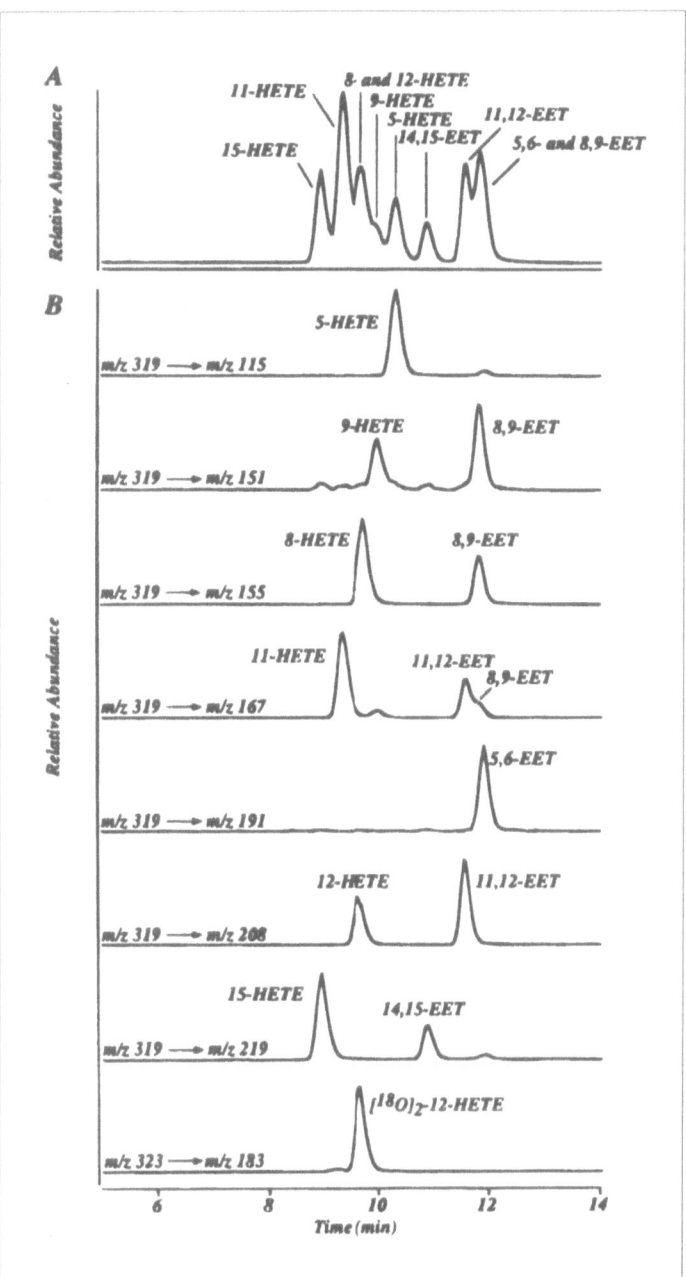

Figure 11
Separation and identification of a series of isomeric epoxy- and hydroxyeicostrienoic acids
by on-line HPLC and tandem MS. Reproduced from [45] with permission.

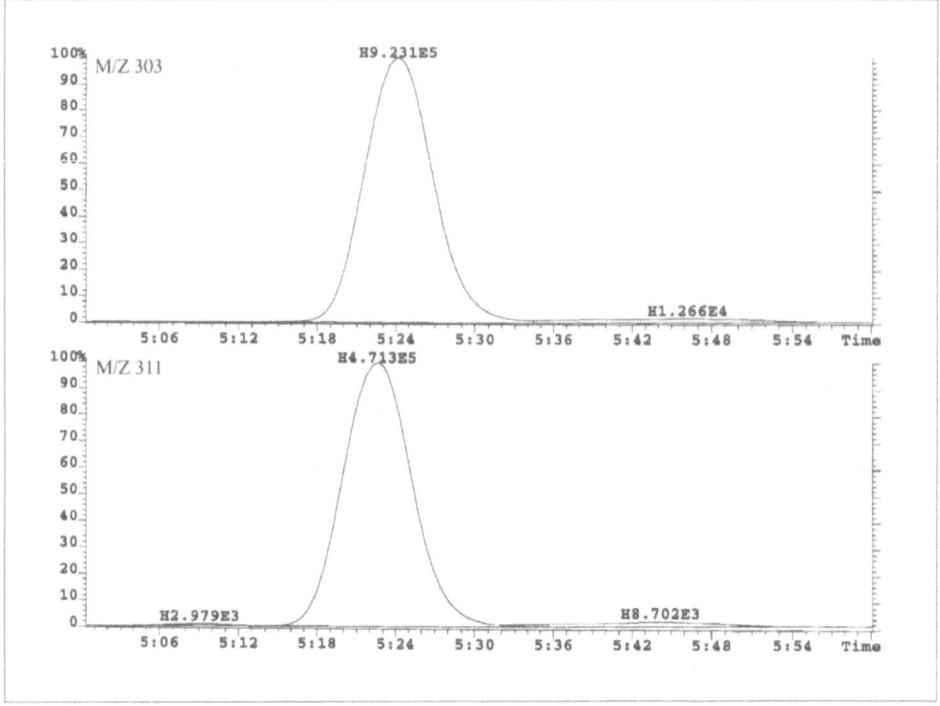

Figure 12
Negative ion GC-MS quantitative analysis of arachidonic acid. Cell culture medium was cen-
trifuged and extracted with heptane, and the dried residue was derivatised with pentafluo-
robenzyl bromide:acetonitrile:diisopropylethylamine. The upper trace is the NICI-MS signal
for m/z 303 (AA), and the lower trace represents the corresponding signal for the internal
standard ([²H₈]-AA) at m/z 311.

concentration of the analyte can be determined from an appropriate calibration set. Thus, the analysis of AA itself can be made using an octa-deuterated homologue, and after extraction and derivatisation, the GC-MS traces for the signal from the endogenous arachidonate and the internal standard are as shown in Figure 12.

The same experiment can be performed using combined HPLC-MS. A straightforward extraction procedure is employed with no derivatisation, and the stable isotope dilution technique permits simultaneous quantitation of AA, prostaglandin E_2 and 12-HETE, extracted from a stimulated keratinocyte culture (Fig. 13) [51].

Immunoassay procedures are also widely available for the quantitative analysis of target lipid mediators [52, 53]. Older procedures have been based on radioim-

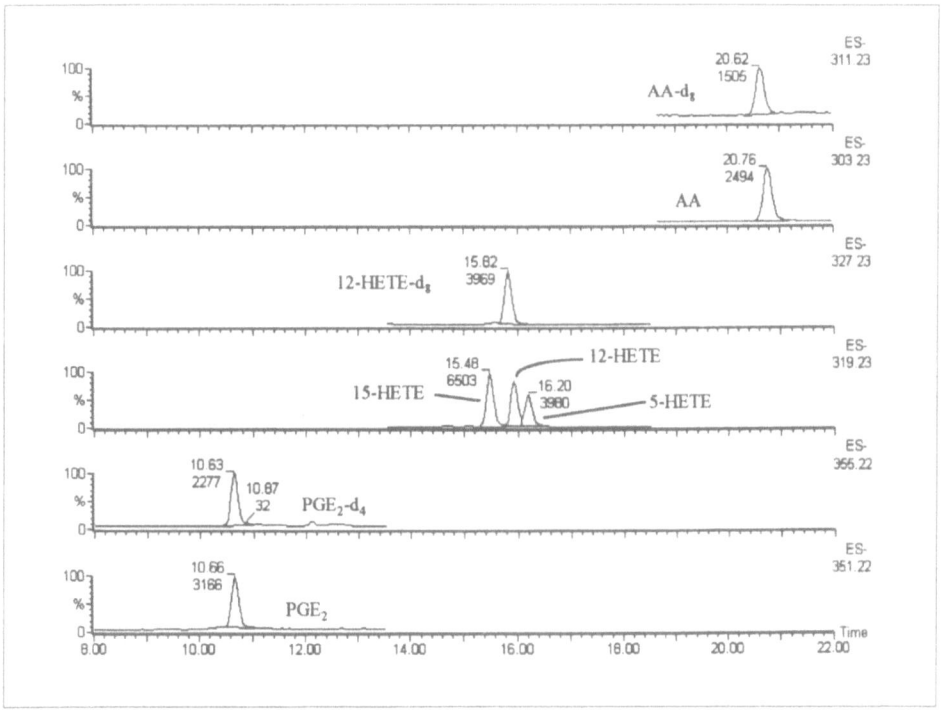

Figure 13
Negative ion electrospray LC-MS of arachidonic acid, three monohydroxyeicosatetraenoic acids and prostaglandin E_2 together with three deuterated internal standards. Cell culture medium was purified on an ODS SPE cartridge and chromatographed on a 15 cm × 1 mm ODS-2 HPLC column with an acetonitrile:aq formic acid gradient. Detection of the [M-H]- ions on a Platform I mass spectrometer in SIM mode.

munoassay methods, but most modern procedures employ enzyme-linked immuno-assay protocols. Colorimetric and fluorometric detection leads to high sensitivities. Commercial kits are available for nearly all AA-derived products, though, crucially, AA itself and other fatty acids cannot be assayed by this method, as no sufficiently specific antibody can be raised against these molecules. The advantage of these methods lies in the ability to perform the assay directly on a biological fluid with little or no prior purification. Large numbers of samples can be processed with semi-automatic instrumentation.

The combination of both techniques forms the basis of the most precise and efficient assay method currently available [54]. This method uses an immunoaffinity

Table 3 - A comparison of the principal features of immunoassay and MS quantitative assay procedures

Immunoassay	MS
Direct analysis on biological matrix	Requires sample extraction. GC-MS requires derivatisation; LC-MS does not.
Specificity depends on antibody. Unforeseen cross reactions will not be evident.	Presence of unexpected compounds will usually be signalled by imperfect chromatography and mass spectrometric behaviour.
Sensitivity to approx. 1–5 pg/ml	Sensitivity varies with compound; 1–10 pg applied to instrument is common.
Multi-well plates permit automatic analysis of large numbers of samples.	Samples have to be processed and analysed one at a time. Automation for this available but expensive.

column to extract the analyte directly from the biological matrix, and the analyte is then washed, eluted with an organic solvent and analysed by GC- or LC-MS with only minimum purification and derivatisation being required. The analysis of prostacyclin, thromboxane and their metabolites is routinely performed by this method in our laboratories [55].

Table 3 shows a summary of comparisons between the two techniques for quantitative analysis in this field. It is probably best to keep MS methods for the cases where no antibody exists and for a gold standard reference assay against which to validate immunoassay-based methods. Where positive identification of a substance is required, MS methods are essential. When limits of detection are the major hurdle in an analysis, it is helpful to remember that the principal losses are always those associated with nonspecific adsorption to vessel walls. The fewer handling operations that can be carried out, the better.

References

1 Christie WW (1982) Lipid analysis, 2nd ed. Pergamon Press, Oxford
2 Hamilton RJ, Hamilton S (eds) (1992) *Lipid analysis: a practical approach*. Oxford University Press, Oxford
3 Baer AN, Costello PB, Green FA (1990) 13(R,S)-HODES free and esterified in psoriatic skin. *J Lipid Res* 131: 125–130

4   Kaluzny MA, Duncan LA, Merritt MV, Epps DE (1985) Rapid separation of lipid class-
    es in high yield and purity using bonded phase columns. *J Lipid Res* 26: 135–140.

5   Henderson RJ, Tocher DR (1992) Thin layer chromatography. In: Hamilton RJ, Hamil-
    ton S (eds): *Lipid analysis: a practical approach*. Oxford University Press, Oxford,
    65–112

6   Folch J, Lees M, Stanley GHS (1975) A simple method for the isolation and purification
    of total lipids from animal tissue. *J Biol Chem* 226: 497–509

7   Bligh EG, Dyer WJ (1959) A rapid method of total lipid extraction and purification.
    Can J Biochem Physiol 37: 911–917

8   Gutnikov G (1995) Fatty acid profiles of lipid samples. *J Chromatogr B* 671: 71–89

9   Mallet AI, Barr RM, Newton JA (1986) Analysis of dihydroxyeicosatetraenoic acids by
    gas chromatography-mass spectrometry. *J Chromatogr B* 378: 194–200

10 Iwase H, Takatori T, Niijima H, Nagao M, Amano T, Iwadate K, Matsuda Y, Nakaji-
    ma M, Kobayashi M (1997) Formation of leukotoxin (9,10-epoxy-12-octadecenoic
    acid) during the auto-oxidation of phospholipids promoted by hemoproteins. *Biochim
    Biophys Acta* 1345: 27–34

11 Schneider C, Schreier P, Herderich M (1997) Analysis of lipoxygenase derived fatty acid
    hydroperoxides by electrospray ionisation tandem mass spectrometry. *Lipids* 32: 331–
    336

12 Schlenk H, Gellerman JL (1960) Esterification of fatty acids with diazomethane on a
    small scale. *Anal Chem* 32: 1412–1414

13 Bannon CD, Craske JD, Hai NT, Harper VL, O'Rourke KL (1982) Analysis of fatty acid
    methyl esters 2. *J Chromatogr* 247: 63–69

14 Morrison WR, Smith LM (1964) Preparation of fatty acid methyl esters and dimethyl
    acetals from lipids with boronfluoride-methanol. *J Lipid Res* 5: 600–608

15 Huang ZH, Wang J, Gage DA, Watson JT, Sweeley CC, Husek P (1993) Characterisa-
    tion of N-ethoxycarbonyl ethyl esters of amino acids by mass spectrometry. *J Chro-
    matogr* 635: 271–281

16 Christie WW (1989) *Gas chromatography of lipids*. Oily Press, Scotland

17 Gopaul NK, Nourooz-Zadeh J, Mallet AI, Anggard EE (1994) Formation of PGF2-iso-
    prostanes during the oxidative modification of low density lipoprotein. *Biochem Bio-
    phys Res Commun* 200: 338–343

18 Blau K, Halket J (eds) (1993) *Handbook of derivatives for chromatography*. John Wiley
    and Sons, Chichester

19 Murphy RC, Sala A (1990) Quantitation of sulphopeptido leukotrienes in biological flu-
    ids by gas chromatography mass spectrometry. *Methods Enzymol* 187: 90–97

20 Woollard PM, Mallet AI (1984) A novel gas chromatographic-mass spectrometric assay
    for monohydroxy fatty acids. *J Chromatogr B* 306: 1–21

21 Knapp D (1979) *Handbook of analytical derivatisation reactions*. John Wiley and Sons,
    Chichester

22 Dell A, Tiller PR (1986) A novel mass spectrometric procedure to rapidly determine the

position of O-acylated residues in the sequence of naturally occurring oligosaccharides. *Biochim Biophys Res Commun* 135: 1126–1134

23   Spivak VA, Shcherbukin VV, Orlov VM, Varsharsky JM (1971) Quantitative ultramicroanalysis of amino acids in the form of their DNS derivatives II. *Anal Biochem* 39: 271–281

24   Gandelman MS (1983) Liquid chromatography detection of cardiac glycosides, saccharides and hydrocortisone based on the photoreduction of 2-t-butylanthroquinone. *Anal Chim Acta* 155: 159–171

25   Evershed RP (1992) Gas chromatography of lipids. In: Hamilton RJ, Hamilton S (eds) *Lipid analysis: a practical approach*. Oxford University Press, Oxford, 113–152

26   Gunstone F (ed) (1996) *Fatty acid and lipid chemistry*. Blackie Academic and Professional, London

27   Patel S, Noble WC (1987) Skin surface lipids. *Br J Dermatol* 117: 735–740

28   Borgeat P, Picard S, Vallerand P, Bourgoin S, Odeimat AS, Sirois P, Poubelle PE (1990) Automated on-line extraction and profiling of lipoxygenase products of arachidonic acid by high performance liquid chromatography. *Methods Enzymol* 187: 98–115

29   Sewell PA (1992) High performance liquid chromatography. In: Hamilton RJ, Hamilton S (eds) *Lipid analysis: a practical approach*. Oxford University Press, Oxford, 153–204

30   Neubert R, Ratyh K, Schiewe J (1997) Capillary zone electrophoresis in skin fatty acid analysis. *Pharmazie* 52: 212–215

31   Olsson NU, Salem N (1997) Molecular species analysis of phospholipids. *J Chromatogr B* 692: 245–256

32   Kim H-Y, Wang T-C L, Ma Y-C (1994) Liquid chromatography/mass spectrometry of phospholipids using electrospray ionisation. *Anal Chem* 66: 3977–3982

33   Schwenk U, Schroder LM (1995) 5-Oxo-eicosanoids are potent eosinophil chemotactic factors. Functional characterisation and structural requirements. *J Biol Chem* 270: 15029–15036

34   Chavis C, Vachier I, Chanez P, Bousquet J, Godard P (1996) 5(S),15(S)-Dihydroxyeicosatetraenoic acid and lipoxin generation in human polymorphonuclear cells: dual specificity of 5-lipoxygenase towards endogenous and exogenous precursors. *J Exp Med* 183: 1633–1643

35   Wheelan P, Zirrolli JA, Morelli JG, Murphy RC (1993) Metabolism of leukotriene B4 by cultured human keratinocytes. *J Biol Chem* 268: 25439–25448

36   Reynaud D, Pace-Asciak CR (1997) Docosahexanoic acid causes accumulation of free arachidonic acid in rat pineal gland and hippocampus to form hepoxilins from both substrates. *Biochim Biophys Acta* 1346: 305–316

37   Murphy, RC (1993) Mass spectrometry of lipids. In: Snyder F (ed): *Handbook of lipid research*, vol. 7. Plenum Press, New York

38   Murphy RC (1995) Lipid mediators, leukotrienes and mass spectrometry. *J Mass Spectrom* 30: 5–16

39   Evershed RP (1992) Mass spectrometry of lipids. In: Hamilton RJ, Hamilton S (eds) *Lipid analysis: a practical approach*. Oxford University Press, Oxford, 263–308

40 Christie WW (1996) Mass spectrometry of fatty acid derivatives. In: Newton RP, Walton TJ (eds): *Application of modern mass spectrometry in plant science research.* Proceedings of the Phytochemical Society of Europe, Vol 40. Oxford University Press, Oxford

41 Harvey DJ (1984) Picolinyl derivatives for the structural determination of fatty acids by mass spectrometry. *Biomed Mass Spectrom* 11: 340–345

42 Tomer KB, Crow FW, Gross ML (1983) Location of double bond position in unsaturated fatty acids by negative ion MS/MS. *J Am Chem Soc* 105: 5487–5488

43 Anton R, Abian J, Vila L (1995) Characterisation of arachidonic acid metabolites through the 12-lipoxygenase pathway in human epidermis by high performance liquid chromatography and gas chromatography mass spectrometry. *J Mass Spectrom* and *Rapid Commun Mass Spectrom* S169–S182

44 Pettitt TR, Rowley AF, Barrow SE, Mallet AI, Secombes CJ (1991) Synthesis of lipoxins and other lipoxygenase products by macrophages from the rainbow trout, *Onchorhynchus mykiss. J Biol Chem* 266: 8720–8726

45 Nakamura T, Bratton D, Murphy RC (1977) Analysis of epoxyeicosatrienoic acids esterified to phospholipids in human red blood cells by electrospray tandem mass spectrometry. *J Mass Spectrom* 32: 888–896

46 Watson AD, Leitinger N, Navab M, Faull KF, Hörkkö S, Witztum JL, Palinski W, Schwenke D, Salomon RG, Sha W et al (1997) Structural identification by mass spectrometry of oxidised phospholipids in minimally oxidised low density lipoprotein that induce monocyte/endothelial interactions and evidence for their presence *in vivo. J Biol Chem* 272: 13579–13607

47 Blair IA (1990) Electron-capture negative-ion chemical ionisation mass spectrometry of lipid mediators. *Methods Enzymol* 187: 13–23

48 Wendelborn DF, Morrow JD, Roberts LJ II (1990) Quantification of $9\alpha,11\beta$-prostaglandin F_2 by stable isotope dilution mass spectrometric assay. *Methods Enzymol* 187: 51–61

49 Turk J, Stump WT, Conrad-Kessel W, Seabold RR, Wolf BA (1990) Quantitation of epoxy and dihydroxyeicosatrienoic acids by stable isotope dilution mass spectrometry. *Methods Enzymol* 187: 175–186

50 Catella F, FitzGerald GA (1990) Measurement of thromboxane metabolites by gas-chromatography-mass spectrometry. *Methods Enzymol* 187: 42–50

51 Newby, C, Mallet, AI. Rapid simultaneous analysis of prostaglandin E_2, 12-hydroxy-eicosatetraenoic acid and arachidonic acid using high performance liquid chromatography-electrospray ionisation mass spectrometry. *Rapid Commun Mass Spectrom* 11: 1723–1727

52 Pradelles P, Antoine C, Lellouche J-P, Maclouf J (1990) Enzyme immunoassays for leukotrienes C_4 and E_4 using acetylcholinesterase. *Methods Enzymol* 187: 82–89

53 Pradelles P, Grasse J, Maclouf J (1990) Enzyme immunoassay of eicosanoids using acetylcholinesterase. *Methods Enzymol* 187: 24–33

54 Vrbanac J, Cox JW, Eller TD, Knapp DR (1990) Immunoaffinity purification-chro-

matographic quantitative analysis of arachidonic acid metabolites. *Methods Enzymol* 187: 62–69

55   Ritter JM, Barrow SE, Doktor HS, Stratton PD, Edwards JS, Henry JA, Gould S (1993) Thromboxane A2 receptor antagonism and synthase inhibition in esential hypertension. *Hypertension* 22: 197–203

The effect of fatty acid composition and retinoic acid on human keratinocyte plasma membrane viscosity

Cynthia L. Marcelo[1] and William R. Dunham[2]

[1]Department of Surgery, University of Michigan Medical School, Kresge I, R-5659, Ann Arbor, MI 48109-0592, USA;
[2]Biophysics Research Division, University of Michigan Medical School, Ann Arbor, MI 48109-1055, USA

Introduction

Lipids, specifically fatty acids and their derivatives, play important and widely varying roles in human epidermis. They are key building blocks in all membranes and in the structures forming the outer layers of skin, and they undergo an active metabolic process that provides many compounds other than the important β-oxidation product, acetyl coenzyme A (CoA). This review focuses on the metabolism of these lipids in mammalian tissues, in the skin and its epidermal component, and in human epidermal cells grown in culture (keratinocytes).

The epidermis is unique in the breadth of differentiation seen in single cell types within one organ. In this tissue compartment, fatty acid synthesis and breakdown is confined to the lower, viable cell layers. Different lipid metabolic pathways have been described for the uppermost differentiating layers and will not be covered here. General references will be used in our discussion and will be cited only once [1, 2]. We will describe an essential fatty acid (EFA)-deficient epidermal cell system that is used to study fatty acid metabolism and will present a possible mechanism for the effect of all-trans retinoic acid on epidermal function.

The fatty acids present in the epidermis are those seen in all other mammalian tissues

Fatty acids are long hydrocarbon (acyl) chains with a carboxyl group at one end. Numbered abbreviations are used to describe the carbon numbers and position of the double bonds in the acyl chain. For example, the saturated fatty acid, stearic acid, having 18 carbons and no double bonds is abbreviated 18:0. Oleic acid has 18 carbons and one double bond. This fatty acid is a monounsaturated fatty acid and is written 18:1 (n-9), with n-9 signifying that the first double bond is at the ninth carbon from the methyl end of the acid. Linoleic acid is another 18-carbon fatty acid with two double bonds, abbreviated 18:2 (n-6) and is a polyunsaturated fatty acid.

Fatty Acids and Inflammatory Skin Diseases, edited by J.-M. Schröder

The n-6 group is the predominant family of polyunsaturated fatty acids in human skin. Arachidonic acid, 20:4(n-6) is formed from 18:2(n-6) (linoleic acid) by the alternating sequence of desaturation and chain elongation via 18:3(n-6) to form 20:3(n-6) fatty acid, to form 20:4(n-6) fatty acid. The n-3 series of polyunsaturated fatty acids is also found in skin. The most abundant n-3 acyl chain in animal tissues is the 22:6(n-3) fatty acid. This fatty acid is found in the membrane phospholipids of cerebral cortex, retina, muscle and several other tissues, but not in skin [1]. Polyunsaturated fatty acids are classified as essential and must be supplied by dietary intake. This is because human cells are unable to desaturate fatty acids beyond the 9-carbon position as required for *de novo* synthesis of the polyunsaturated fatty acids.

Fatty acids are usually esterified to the 1- and 2-positions of the glycerol backbone of phospholipids and acylglycerides. The phospholipids include phosphatidylcholine, phosphatidylethanolamine, phosphatidylserine, phosphatidylinositol and cardiolipin and are the major class of lipids found in mammalian cell membranes. The acylglycerides function as the fat stores for the cell.

In epidermis, the predominant fatty acids are the saturated fatty acids: myristic acid, 14:0; palmitic acid, 16:0; stearic acid, 18:0; arachidic acid, 20:0; behenic acid, 22:0; and lignoceric acid, 24:0; the monounsaturated fatty acids: palmitoleic acid, 16:1 (n-7); oleic acid, 18:1 (n-9); and the polyunsaturated fatty acids: linoleic acid, 18:2 (n-6); dihomo-γ-linoleic acid, 20:3 (n-6); and arachidonic acid, 20:4 (n-6) [3, 4] (see Tab. 1).

Fatty acids are carried to the epidermis either as free fatty acids complexed to serum albumin or as triglycerides associated with lipoproteins, from which they diffuse though the cell membrane. Inside the cell, fatty acids are not usually found as free acids. They are esterified to the phospholipids and acylglycerides, are in the "active" thioester of fatty acyl-CoA, or are attached to various fatty acid-binding proteins. These fatty acid-binding proteins were initially thought to mediate the transport of fatty acids into the epidermal basal cell. Now it is known that they function as temporary storage molecules, thus limiting the amount of free acids and long-chain fatty acyl-CoAs in the cell cytoplasm. Fatty acid-binding proteins also transport fatty acids among the differentiating compartments in the upper layers of the epidermis [5–9].

The release of fatty acids esterified to phospholipids in the cell membrane is mediated predominantly by two enzymes. The first, phospholipase A_2 (PLA$_2$) was originally identified in skin by Long and Yardley in 1972 [10]. This, mostly cytosolic, enzyme was purified from epidermis [11], and a very minor microsomal PLA$_2$ has been identified using molecular probes [12, 13]. The second enzyme is phospholipase C (PLC), identified as PLC-γ-1. Investigators have proposed that cells, including keratinocytes have multiple mechanisms to regulate this lipase. Several of these second messengers are calcium, transforming growth factor-α (TGFα), epidermal growth factor (EGF) and the tumor promoter, phorbol myristate acetate

Table 1 – Representative fatty acid levels obtained from adult human skin biopsies[1] and from EFAD and normalized cell cultures[2] (bold face figures highlight most important changes)

fatty acid	epidermal biopsy	EFAD cells	Normalized cells[3]
14:0	1.1	2.3	2.1
16:0	16.4	16.6	18.0
16:1	–	15.5	5.1
18:0	12.9	12.5	13.4
18:1	**16.3**	**41.5**	**15.4**
18:2	**25.3**	**4.1**	**27.1**
20:0	1.2	–	–
20:2	–	–	1.1
20:3	2.1	1.3	9.6
20:4	**8.3**	**3.4**	**8.3**
22:0	3.8	–	–
24:0	12.5	1.6	–
24:1	–	1.0	–

[1]Values obtained by choosing the fatty acid components greater than 1%, then normalizing the sum to 100. Data available in [65].
[2]Values obtained by choosing the fatty acid components greater than 1%, then normalizing the sum to 100 [66].
[3]Normalized cells refer to cell cultured in 2×/1× medium containing 10 μM 18:2 + 5 μM 16:0 + 5 μM 20:4.

(PMA) [14–16]. The released fatty acids are activated by conversion to thioesters such that they can enter the primary catabolic pathway of fatty acids, β-oxidation.

Catabolism of fatty acids via β-oxidation occurs in mitochondria and peroxisomes

Fatty acid thiokinase enzyme is named adenosine triphosphate (ATP)-dependent acyl-CoA synthetase and has nicotinamide adenine dinucleotide (NAD) and CoA (CoASH) as cofactors. There are at least three kinds of fatty acid synthetases: a short-chain synthetase using predominantly acetate (C2) as substrate, a medium-chain synthetase acting on both saturated and unsaturated C4 to C12 chain length fatty acids, and long-chain synthetase activating C14 to C22 saturated and unsaturated fatty acids. All three are particulate enzymes localized in cellular membranes, with the short- and medium-chain synthetases found on outer mitochondrial membranes. The long-chain enzyme is found in the endoplasmic reticulum. After activation, the fatty acyl-CoA is transported across the mitochondrial membrane via carnitine-dependent uptake and translocation (the acyl-carnitine transport system).

Once inside the mitochrondrion, the fatty acyl-CoA is degraded via four enzymatic actions: dehydrogenation, hydration, oxidation and thiolytic cleavage. The shortened fatty acyl-CoA is again recycled through these four reactions, resulting in a number of acetyl-CoA units corresponding to the length of the original fatty acid. For example, 18:0 would yield 9 $CH_3-C=O-SCoA$ units, which upon entering the mitochondrial citric acid cycle (TCA cycle) yield 18 CO_2 + 18 H_2O + 147 ATP. Unsaturated fatty acids undergo this sequence of β-oxidation with a few additional steps, requiring participation of special isomerases, hydrases and epimerases.

Another type of β-oxidation occurs in special structures called peroxisomes. These reactions use a different set of protein enzymes to shorten long-chain fatty acids to medium-chain fatty acids [17–20]. A number of diseases have been identified that have inborn errors of the very long chain synthetase [9, 21–27]. Abnormal fatty acid metabolism of fatty acids is seen in dermal fibroblasts and in the epidermal component of persons with this disease, indicating that peroxisomal β-oxidation is an important pathway in skin.

Anabolism of fatty acids occurs via *de novo* synthesis, and elongation and saturation reactions

Fatty acids are synthesized *de novo* in the cytoplasmic component of the cell. The decarboxylation of pyruvate to acetyl-CoA via the TCA cycle in the mitochondrion supplies the substrates for the *de novo* synthesis of fatty acids. Acetyl-CoA leaves the mitochrondrion via the acyl-carnitine transport system or via the citrate-cleaving enzyme mechanism. Once in the cytoplasm the acetyl-CoA plus CO_2 generated by other cell metabolic pathways are acted on by acetyl-CoA carboxylase in a biotin- and ATP-dependent reaction to produce malonyl-CoA, the immediate substrate of the fatty acid synthase.

Acetyl-CoA carboxylase is inactivated via hormonal control when phosphorylated by adenosine 5'-monophosphate (AMP)-activated protein kinase and to a lesser degree by cyclic AMP-dependent protein kinase [28–30]. Free fatty acids [31] and palmityl-CoA inactivate this enzyme, a mechanism most probably functional in the epidermis and in culture keratinocytes. In animal models, disruption of the epidermal barrier activates this acetyl-CoA carboxylase as well as CoA synthase, showing that in the epidermis these enzymes are a regulatory control point for *de novo* fatty acid synthesis [32].

Using malonyl-CoA as a substrate, CoA synthase is a multienzyme complex that adds two carbon units as $-CH_3CH_2-$ to build the final product of *de novo* synthesis, 16:0 (palmitic acid). This reaction is NADPH (H) dependent, with the pentose phosphate pathway the most likely energy source for this reaction. Thus 8 acetyl-CoA + 7 ATP + 14 NADPH(H) in the presence of CO_2 yields ADP + $7P_i$ + 8 CoASH + 6 H_2O. The product 16:0 is then activated via thiokinase to form palmityl-CoA that goes mainly into acyl-glycerols and phospholipids. The palmityl-CoA can be transferred into the mitochondrion via the carnitine transport system to synthesize long-chain fatty acids using acetyl-CoA as the condensing unit.

Long-chain fatty acids, both saturated and unsaturated, can also be synthesized via a series of elongation and desaturation reactions occurring in the endoplasmic reticulum. If the fatty acid is free, fatty-CoA ligases form the fatty acyl-CoA. In these reactions, malonyl-CoA is the two-carbon condensing unit. In humans, 16:0 can be acted on by Δ^9 desaturase to form 16:1. Palmitate (16:0) can also, via elongation reactions, form 18:0 that in turn can be acted on by Δ^9 desaturase to form 18:1. Mammals do not posses the enzymes to insert a double bond beyond the ninth carbon, thus 18:2, the first fatty acid in the polyunsaturated group, must be supplied in the diet.

Δ^6 desaturase converts 18:2 to 18:3(n-6) which is converted via elongation to 20:3(n-6) and finally to 20:4(n-6) via a Δ^5 desaturase. These last series of reactions have not been identified in extracts of whole epidermis [33, 34], although they are expressed in *in vitro* systems [35, 36]. We have reported this conversion and measured the *in vivo* kinetic rates of the main constituents in the *de novo* and EFA pathways by simultaneously solving the relevant differential equations [37]. The parameters of these solutions were changed iteratively until the kinetic rate constants matched the time course of fatty acid composition after the addition of radiolabeled acetate, 16:0 or 18:2, to the medium of the EFA cell strains. These experiments show that 18:2 is very rapidly converted to 20:4 in EFA-deficient cells; the characteristic time for the conversion is less than 1 h. The formation of 16:0 from acetate (*de novo* synthesis) is at a similar rate, with subsequent steps much slower.

Fatty acids control epidermal keratinocyte function

Human keratinocytes grown in low calcium medium, free of serum and lipid, show large changes in cellular fatty acid composition [4]. These cells have decreased amounts of n-6 polyunsaturated fatty acids (18:2, 20:3 and 20:4 fatty acids). These are replaced with increased amounts of the monounsaturated fatty acids, 16:1(n-7) and 18:1(n-9), so that these cells are severely EFA-deficient. They grow rapidly, and the primary cultures can be subcultured until passage 6, when senescent cells begin to overtake the culture [38–40].

The EFA-deficient state of the keratinocytes is reflected in the membrane phospholipid composition [41, 42]. Moreover, the fatty acid composition of these cells is controlled by medium fatty acid supplementation [43]. When brought to human epidermis fatty acid levels, the cultures have a slowed growth rate, a decrease in successful passage and a more differentiated (flat and cornified) cellular appearance. The fatty acid composition of both human biopsies and cultured cells is presented in Table 1, which will later be considered in more detail.

In vitro EFA-deficient keratinocyte cultures are used to study the effect of fatty acids on cell function and metabolism

Keratinocytes cultured in basic medium MCDB 153 prepared as described by Boyce and Ham [44] are extremely EFA-deficient [4]. Fatty acid supplementation can be accomplished by feeding the cells with fatty acid-supplemented medium at the P0 to P1 passage. The "supplemented" media are: 2× = 10 µM 18:2 (n-6) and 5 µM 16:0; 2×/1× = 10 µM 18:2 (n-6), and 5 µM 16:0 and 20:4 (n-6); oleic = 10 µM 18:1 (n-7) and 5 µM 16:0. These media restore the fatty acid profile of these cells to approximately those seen in epidermal biopsies. Table 1 presents the molar percentage of the major fatty acids found in the epidermis. The EFA-deficient cells show a large increase in the content of 16:1 and 18:1 fatty acid when compared with the biopsy material, while the amount of 18:2 fatty acid is greatly diminished and that of 20:4 fatty acid is halved. Normalization of the cellular fatty acids with medium supplemented with both 18:2 and 20:4 fatty acid, designated 2×/1× medium, restores the 18:2, 20:4 and 18:1 fatty acid values of the EFA-deficient cells to those of epidermal tissue. One anomaly of EFA-fatty acid medium supplementation is that a larger than normal amount of 20:3 fatty acid accumulates in the cells.

Membrane viscosity is determined in adult human keratinocytes using electron paramagnetic resonance (EPR) spectroscopy. We have measured the viscosity of these cells as documented in Dunham et al. [45]. The spin label was 20 µM 16-doxyl stearate methyl ester and was delivered to the cells in monolayer at 4° C by 12 µM fatty acid-free bovine serum albumin for 15 min. Data generated by this analytical system was used to predict the viscosity of the keratinocyte membrane using the following equation:

$$\eta = 69.4 + 0.45\,[16{:}1] + 0.97\,[18{:}1] - 0.71\,[18{:}2] - 0.28\,[20{:}3] \qquad [1]$$

where the fatty acid concentrations are in mole percent and the viscosity is in cP. This equation was empirically derived from measurements of keratinocytes with membrane fatty acid compositions altered by growth in media supplemented with EFAs [45]. As reported by us, these cells show a factor of 2 modification of plasma membrane viscosity due to changes in the EFA supplementation of the medium [46].

All-trans retinoic acid controls keratinocyte growth and fatty acid composition

All-trans retinoic acid is a lipid modifier of keratinocyte growth and differentiation. The use of all-trans retinoic acid in the therapy of skin diseases is well documented. *In vitro* this retinoid inhibits keratinocyte proliferation when the cells are grown in serum-containing medium. The effect is contrary to that seen in whole skin, where stimulation of proliferation is seen [47–50]. Both effects required 0.5–2.0 µM all-trans retinoic acid, which is approximately 50- to 200-fold higher than is necessary to activate the nuclear retinoic acid receptors (RARs) [51, 52].

We have reported that 0.5 and 1 µM all-trans retinoic acid prevented the drastically decreased cell growth and loss of cell viability, and increased cell senescence induced by EFA supplementation. At both concentrations all-trans retinoic acid was seen to alter the fatty acid content of the EFA-supplemented cultures, the largest change being in the 16:1 and 18:2 fatty acids [53]. Similar effects of retinoids on epidermal lipids have been reported by Jensen et al. [54], Ponec et al. [55] and in retinoic acid receptor-defective transgenic mice [56, 57].

We have also shown that in keratinocyte membranes the EFA and non-EFA content of the phospholipids correlated with plasma membrane viscosity as measured by electron paramagnetic resonance spectroscopy [46]. The modification of membrane fatty acids induced by all-trans retinoic suggested that the viscosity of the cell membrane was also changed by this retinoid. We investigated this possibility by measuring the effect of all-trans retinoic acid on EFA-deficient and EFA-supplemented cell membrane viscosity using spin probes and EPR spectroscopy.

Growth of keratinocytes in 1.0 µM all-trans retinoic acid lowers the membrane viscosity of keratinocytes and changes the fatty acid composition of the membrane

Figure 1 presents the membrane viscosity values as measured by EPR in keratinocytes grown in various fatty acid-supplemented media. The cells were exposed to all-trans retinoic acid for 4 to 7 days. All-trans retinoic acid lowered the measured membrane viscosity of cells grown in every media formulation. For each medium, the difference between no retinoic acid (clear bar) and 1.0 µM retinoic acid (solid bar) was significant at p ≤ 0.05.

As presented in Table 2, 1 µM all-trans retinoic acid changed the levels of a number of fatty acids in keratinocytes grown in control and fatty acid-supplemented media. Increases in 14:0, 16:1 and 18:1 fatty acid occurred, while the amount of 18:2 fatty acid decreased. The difference in measured viscosity between the cells grown in the presence and absence of all-trans retinoic acid is also presented in Table 2. Statistically significant differences of –17 to –27 cP were measured between

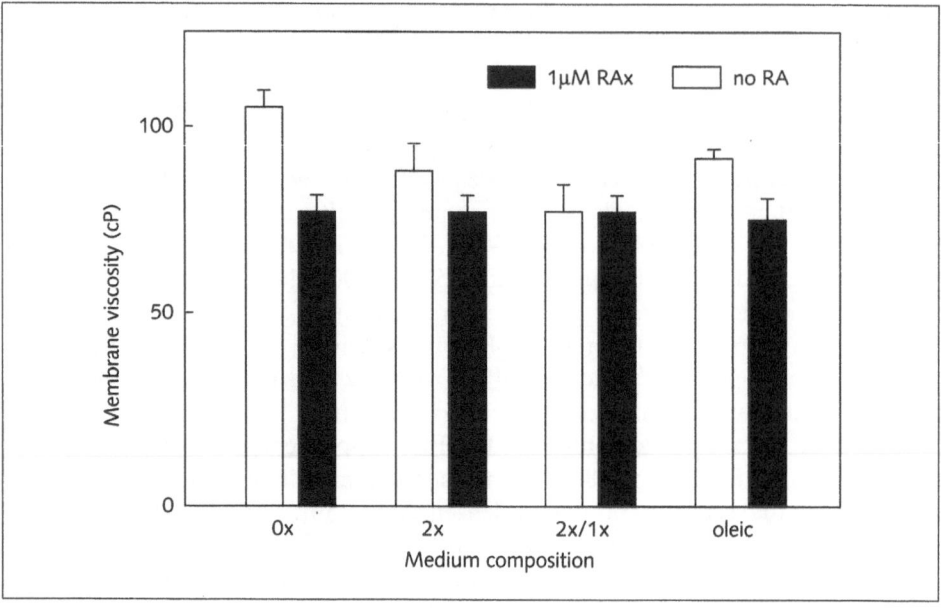

Figure 1

All-trans retinoic acid decreases the membrane viscosity of adult human keratinocytes in cul-
ture. The cells were grown in four media: 0x = EFA-deficient medium, n = 14; 2x = 10 μM
18:2 and 5 μM 16:0, n = 8; 2x/1x = 10 μM 18:2 and 5 μM 16:0 and 20:4, n = 12; oleic =
10 μM 18:1 and 5 μM 16: 0, n = 5. One μM all-trans was added to the cultures 4–7 days
prior to EPR determination of membrane viscosity using 16-doxyl-stearate methyl ester as
the spin probe. All differences beween "no RA" and "1 mM RA" are statistically significant
at p ≤ 0.5.

the cells with and without all-trans retinoic acid. Thus, all-trans retinoic acid
decreased the viscosity, that is, increased the fluidity, of the cell membranes.

The predicted viscosity of the cell membrane, obtained by substituting into
Equation 1 using the fatty acid values presented in Table 2, showed that the cells
should have had no change or a slight increase in viscosity.

All-trans retinoic acid affects cell fatty acid composition and viscosity at both 0.5 and 1.0 μM concentrations

All-trans retinoic acid at 0.5 μM changed the fatty acid composition of the cells as
well as the measured viscosity (Table 3). Again, no change or a slight increase in viscosity would have been predicted using Equation 1.

Table 2 - Changes in percent fatty acid composition due to 1 µM retinoic acid addition to the culture media[1]

Fatty acid	Control	2×	2×/1×	2× (oleic)
14:0	0.2	1.6[2]	1.0*	−0.1
16:0	2.2[2]	2.5[2]	2.9	−1.8
16:1	1.5	6.0[2]	2.5	3.7[2]
18:0	−0.7	−3.0[2]	−3.0[2]	−1.7[2]
18:1	−2.6[2]	0.9	1.6	−1.5
18:2	0.1	−8.1[2]	−6.1[2]	0.5
20:3	−0.2	−0.2	0.7	−0.1
Measured viscosity	−27 cP[2]	−22 cP[2]	−20 cP[2]	−17 cP[2]
Calculated viscosity[3]	−2 cP	9 cP	7 cP	0 cP
N	14	8	12	5

[1]Differences are between cultures grown with and without all-trans retinoic acid. 2× = 10 µM 18:2 + 5 µM 16:0; 2x/1x = 10 µM 18:2 + 5 µM 16:0 + 5 µM 20:4; 2x (oleic) =10 µM 18:1 + 5 µM 16:0.
[2]$p < 0.05$, paired Student's t-test, two-tailed.
[3]$\eta = 69.4 + 0.45\ [16:1] + 0.97\ [18:1] − 0.71\ [18:2] − 0.28\ [20:3]$ (Eq. 1).

Table 3 - Changes in percent fatty acid composition due to 0.5 µM or 1.0 µM retinoic acid addition to the culture media[1]

Fatty acid	0.5 µM RA	1.0 µM RA
14:0	1.3[2]	1.3[2]
16:0	2.2[2]	2.8[2]
16:1	3.9[2]	4.0[2]
18:0	−2.5[2]	−2.9[2]
18:1	0.7	1.1
18:2	−7.1[2]	−7.4[2]
20:3	1.0	1.2[2]
Measured viscosity	−16 cP[2]	−18 cP[2]
Calculated viscosity[3]	7 cP	8 cP
N	10	10

[1]RA, all-trans retinoic acid. Differences are between cultures grown with and without all-trans retinoic acid.
[2]$p < 0.05$, paired Student's t-test, two-tailed.
[3]$\eta = 69.4 + 0.45\ [16:1] + 0.97\ [18:1] − 0.71\ [18:2] − 0.28\ [20:3]$ (Eq. 1).

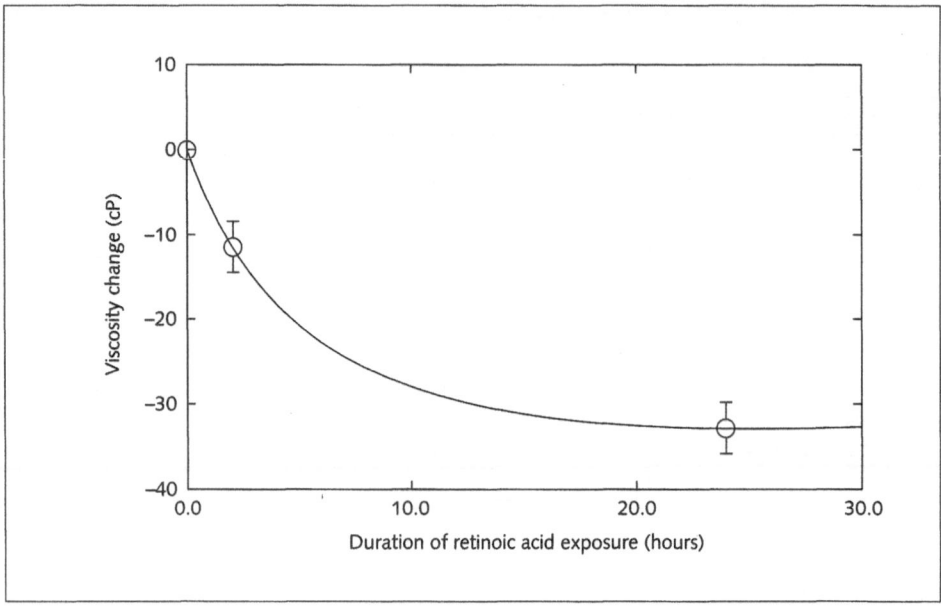

Figure 2
All-trans retinoic acid directly decreases keratinocyte membrane viscosity. Keratinocytes maintained at 4°C were exposed to all-trans retinoic acid for the times indicated, after which EPR measurements were made of viscosity. At the 2-h time point, a significant decrease in membrane viscosity was measured in cells kept 4°C to halt cell chemistry and cell signaling. The solid line is an exponential decay fit to the data ($t_{1/2}$ = 3.5 h).

All-trans retinoic acid decreases the viscosity of cells 2 h after addition

Keratinocytes maintained at 4°C to halt cell chemistry and signaling were exposed to 1.0 µM all-trans retinoic acid for 2 h. EPR analysis showed that the viscosity decreased 11 ± 3 cP over the 2-h time period. This difference was significant at $p \leq 0.5$, n = 11. This viscosity value, along with a 24-h viscosity change of –33 cP ± 3 (n = 11) was used to generate a graph modeling the time evolution of the viscosity (Fig. 2). In this graph, the assumption was made that the viscosity approached its final asymptote exponentially. This assumption was not justified experimentally (except by the data in Fig. 2), but does allow us to characterize the rate of the viscosity change by a single number. In this case, $t_{1/2}$ was determined to be 3.5 h. Thus, all-trans retinoic acid can alter cell membrane viscosity by physically entering the lipid component of the membrane. This effect is near completion by 10 h (94% complete).

Conclusion

The information covered in this review suggests that cell membrane viscosity, the effect of lipid modulators of keratinocyte function such as all-trans retinoic acid and fatty acid metabolism are integrally coupled.

The acid form of vitamin A, all-trans retinoic acid, affects the growth and differentiation of both the dermal and epidermal components of skin, *in vivo* and *in vitro* [58–64]. The extent and direction of the effect vary according to the model system and are presumed to result from undefined differences in the "conditions" of the growth system [58].

The data presented in this report show that all-trans retinoic acid decreased the measured membrane viscosity of all the keratinocytes irrespective of whether the cells were EFA-deficient or EFA-supplemented. Additionally, short-term loading of the cell membranes with all-trans retinoic acid showed (Fig. 2) that this retinoid lowered the viscosity of the cells during a 2-h short exposure at 4° C, and that by approximately 10 h, the retinoic acid effect on viscosity should be maximal. This suggests that retinoic acid lowers plasma membrane viscosity by acting directly on the membrane and that changes in cell membrane fatty acid composition occurred at a later time point.

These studies support the hypothesis that the effect of micromolar all-trans retinoic acid on keratinocyte function occurs via an initial decrease in cell membrane viscosity, followed by changes in the fatty acid profile of the cells. A viable hypothesis is that all-trans retinoic initially alters the physical environment of the cell membrane as defined by our viscosity measurements. This change then activates or allows expression of metabolic pathways that alter the final membrane fatty acid composition. Alternatively, all-trans retinoic acid could first cause expression of mechanisms at the gene level. This could then result in stimulation of cell growth, inhibition of cell senescence and altered fatty acid metabolism with accompanying alterations in cell membrane fatty acid composition.

Acknowledgments
This work was supported in part by PHS awards AR26009 and AR 42223.

References

1   Vance D, Vance J (eds) (1985) *Biochemistry of lipids and membranes*. Benjamin/Cummings, Menlo Park, CA
2   Bohinski R (ed) (1976) *Modern concepts in biochemistry*. Allyn and Bacon, Boston
3   Chapkin RS, Ziboh VA, Marcelo CL, Voorhees JJ (1986) Metabolism of essential fatty

acids by human epidermal enzyme preparations: evidence of chain elongation. *J Lipid Res* 27: 945–954

4   Marcelo CL, Duell EA, Rhodes, LM, Dunham WR (1992) An *in vitro* model of essential fatty aid deficiency. *J Invest Dermatol* 99: 703–708

5   Street JM, Johnson DW, Singh H, Poulos A (1989) Metabolism of saturated and polyunsaturated fatty acids by normal and Zellweger syndrome skin fibroblasts. *Biochem J* 260: 647–655

6   Madsen P, Rasmussen HH, Leffers H, Honore B, Celis JE (1992) Molecular cloning and expression of a novel keratinocyte protein [psoriasis-associated fatty acid-binding protein (PA-FABP)] that is highly up-regulated in psoriatic skin and that shares similarity to fatty acid-binding proteins. *J Invest Dermatol* 99: 299–305

7   Schurer NY, Bass NM, Jin S, Manning JA, Pillai S, Williams ML (1993) High-affinity fatty acid-binding activity in epidermis and cultured keratinocytes is attributable to high-molecular-weight and not low-molecular-weight fatty acid-binding proteins. *J Invest Dermatol* 100: 82–86

8   Siegenthaler G, Hotz R, Chatellard-Gruaz D, Didierjean L, Hellman U, Saurat JH (1994) Purification and characterization of the human epidermal fatty acid-binding protein: localization during epidermal cell differentiation *in vivo* and *in vitro*. *Biochem J* 302: 363–371

9   Masouye I, Saurat JH, Siegenthaler G (1996) Epidermal fatty-acid-binding protein in psoriasis, basal and squamous cell carcinomas: an immunohistological study. *Dermatology* 192: 208–213

10   Long VJ, Yardley HJ (1972) Phospholipase A activity in the epidermis. *J Invest Dermatol.* 58: 148–154

11   Finnen MJ, Lovell CR (1991) Purification and characterization of phospholipase A_2 from human epidermis. *Biochem Soc Trans* 19: 91s

12   Bergers M, Verhagen B, Jongeruis M, van der Kerkhof P (1988) A unique phospholipase A_2 in human epidermis: its physiologic function and its level in certain dermatoses. *J Invest Dermatol* 90: 23–25

13   Andersen S, Sjursen W, Laegreid A, Volden G, Johansen B (1994) Elevated expression of human nonpancreatic phospholipase A_2 in psoriatic tissue. *Inflammation* 18: 1–12

14   Kast R, Furstenberger G, Marks F (1991) Activation of a keratinocyte phospholipase A2 by bradykinin and beta-phorbol 12-myristate 13-acetate. Evidence for a receptor-GTP-binding protein versus a protein kinase C mediated mechanism. *Eur J Biochem* 202: 941–950

15   Nanney LB, Gates R, Todderud G, King L, Carpenter G (1992) Altered distribution of phospholipase C gamma-1 in benign hyperproliferative diseases. *Cell Growth Differ* 3: 233–239

16   Punnonen K, Denning M, Lee E, Li L, Rhee S, Yuspa S (1993) Keratinocyte differentiation is associated with changes in the expression of phospholipase C isozymes. *J Invest Dermatol* 101: 719–726

17   Singh H, Derwas N, Poulos A (1987) Beta-oxidation of very-long-chain fatty acids and

their coenzyme A derivatives by human skin fibroblasts. *Arch Biochem Biophys* 254: 526–533

18 Jakobs BS, Wanders RJ (1991) Conclusive evidence that very-long-chain fatty acids are oxidized exclusively in peroxisomes in human skin fibroblasts. *Biochem Biophys Res Commun* 178: 842–847

19 Singh H, Brogan M, Johnson D, Poulos A (1992) Peroxisomal beta-oxidation of branched chain fatty acids in human skin fibroblasts. *J Lipid Res* 33: 1597–1605

20 Schmidt A, Vogel RL, Witherup KM, Rutledge SJ, Pitzenberger SM, Adam M, Rodan GA (1996) Identification of fatty acid methyl ester as naturally occurring transcription-al regulators of the members of the peroxisome proliferator-activated receptor family. *Lipids* 31: 1115–1124

21 Singh H, Derwas N, Poulos A (1987) Very long chain fatty acid beta-oxidation by sub-cellular fractions of normal and Zellweger syndrome skin fibroblasts. *Arch Biochem Biophys* 257: 302–314

22 Wanders RJ, van Roermund CW, van Wijland MJ, Schutgens RB, Heikoop J, van den Bosch H, Schram AW, Tager JM (1987) Peroxisomal fatty acid beta-oxidation in rela-tion to the accumulation of very long chain fatty acids in cultured skin fibroblasts from patients with Zellweger syndrome and other peroxisomal disorders. *J Clin Invest* 80: 1778–1783

23 Christensen E, Hagve TA, Christophersen BO (1988) The Zellweger syndrome: deficient chain-shortening of erucic acid [22: 1 (n-9)] and adrenic acid [22: 4 (n-6)] in cultured skin fibroblasts. *Biochim Biophys Acta* 959: 134–142

24 Lazo O, Contreras M, Bhushan A, Stanley W, Singh I (1989) Adrenoleukodystrophy: impaired oxidation of fatty acids due to peroxisomal lignoceroyl-CoA ligase deficiency. *Arch Biochem Biophys* 270: 722–728

25 Raza H, Chung WL, Mukhtar H (1991) Specific high-affinity binding of fatty acids to epidermal cytosolic proteins. *J Invest Dermatol* 97: 323–326

26 Watanabe R, Fujii H, Odani S, Sakakibara J, Yamamoto A, Ito M, Ono T (1994) Mol-ecular cloning of a cDNA encoding a novel fatty acid-binding protein from rat skin. *Biochem Biophys Res Commun* 200: 253–259

27 Jakobs BS, Wanders RJ (1996) Impaired peroxisomal fatty acid oxidation in human skin fibroblasts with a mitochondrial acylcarnitine/carnitine translocase deficiency. *J Inherit Metab Dis* 19: 185–187

28 Faergeman NJ, Knudsen J (1997) Role of long-chain fatty acyl-CoA esters in the regu-lation of metabolism and in cell signaling. *Biochem J* 323: 1–12

29 Winder WW,. Wilson HA, Hardie DG, Rasmussen BB, Hutber CA, Call GB, Clayton RD, Conley LM, Yoon S, Zhou B (1997) Phosphorylation of rat muscle acetyl-CoA car-boxylase by AMP-activated protein kinase and protein kinase A. *J Applied Physiol* 82: 219–225

30 Velasco G, Geelen MJ, Guzman M (1997) Control of hepatic fatty acid oxidation by 5'-AMP-activated protein kinase involves a malonyl-CoA-dependent and a malonyl-CoA-independent mechanism. *Arch Biochem Biophys* 337: 169–175

31 Jacobs RA, Majerus PW (1973) The regulation of fatty acid synthesis in human skin fibroblasts. Inhibition of fatty acid synthesis by free fatty acids. *J Biol Chem* 248: 8392–8401

32 Ottey KA, Wood LC, Grunfeld C, Elias PM, Feingold KR (1995) Cutaneous permeability barrier disruption increases fatty acid synthetic enzyme activity in the epidermis of hairless mice. *J Invest Dermatol* 104: 401–404

33 Chapkin RS, Ziboh VA (1984) Inability of skin enzyme preparations to biosynthesize arachidonic acid from linoleic acid. *Biochem Biophys Res Commun* 124: 784–792

34 Ziboh VA, Chapkin RS (1988) Metabolism and function of skin lipids. *Prog Lipid Res* 27: 81–105

35 Madison KC, Wertz PW, Strauss JS, Downing DT (1986) Lipid composition of culture murine keratinocytes. *J Invest Dermatol* 87: 253–259

36 Isserof RR, Ziboh VA, Chapkin RS, Martinez DT (1987) Conversion of linoleic acid into arachidonic acid by cultured murine and human keratinocytes. *J Lipid Res* 13: 458–467

37 Marcelo CL, Dunham WR (1993) Fatty acid metabolism studies of human epidermal cell culturers. *J Lipid Res* 34: 2077–2090

38 Boyce ST, Ham RG (1985) Cultivation, frozen storage and clonal growth of normal epidermal keratinocytes in serum-free medium. *J Tiss Cult Meth* 9: 83–93

39 Willie JJ, Pittelkow MR, Shipley G.D, Scott RE (1984) Integrated control of growth and differentiation of normal human prokeratinocytes cultured in serum-free medium: clonal analyses, growth kinetics and cell cycle studies. *J Cell Physiol* 121: 31–44

40 Boyce ST, Ham RG (1983) Calcium-regulated differentiation of normal human epidermal keratinocytes in chemically defined clonal culture and serum-free serial culture. *J Invest Dermatol* 81 (Suppl): 33s–40s

41 Williams M, Rutherford SL, Ponec M, Placzek DR, Elias P (1988) Density dependent variation in the lipid content and metabolism of cultured human keratinocytes. *J Invest Dermatol* 91: 86–91

42 Cullis, PR, Hope MJ (1985) Physical properties and functional roles of lipids in membranes. In: Vance D, Vance J (eds): *Biochemistry of lipids and membranes*. Benjamin/Cummings, Menlo Park, CA, 29–30

43 Marcelo CL, Rhodes LM, Dunham WR (1994) Normalization of essential-fatty-acid-deficient keratinocytes requires palmitic acid. *J Invest Dermatol* 103: 564–568

44 Boyce ST, Ham RG (1985) Normal human epidermal keratinocytes. In: Weber MM, Sekely L (eds): In vitro *models for cancer research*. CRC Press, Boca Raton, FL, 245–274

45 Dunham WR, Sands RH, Klein SB, Duell EA, Rhodes LM, Marcelo CL (1996). EPR measurements showing that plasma membrane viscosity can vary from 30 to 100 cP in human cell strains. *Spectrochemica Acta A* 52: 1357–1368

46 Dunham WR, Klein S, Rhodes LM Marcelo CL (1996) Oleic acid and linoleic acid are the major determinants of plasma membrane viscosity. *J Invest Dermatol* 107: 332–335

47 Marcelo CL, Madison K (1984) Regulation of the expression of epidermal keratinocyte

proliferation and differentiation by Vitamin A analogs. *Arch Dermatol Res* 276: 381–389

48   Glick AB, Flanders KC, Danielpour D, Yuspa SH, Sporn M (1991) Retinoic acid induces transforming growth factor-beta 2 in cultured keratinocytes and mouse epidermis. *Cell Regulation* 1: 87–97

49   McGuire J, Fedarko N, Johanssen E, La Vigna J, Lyons G, Milstone L, Osber M (1982) The influence of retinoids on cultivated keratinocytes. *J Am Acad Dermatol* 6: 630–639

50   Weiss J, Ellis C, Headington K, Tincoff T, Hamiliton T, Voorhees J (1988) Topical tretinoin improves photoaged skin. A double-blind vehicle-controlled study. *J Am Med Assoc* 259: 527–532

51   Duell E, Anders A, Griffiths C, Chambon P, Voorhees J (1992) Human skin levels of retinoic acid and cytochrome P-450-derived 4-hydroxyretinoic acid after topical application of retinoic acid *in vivo* compared to concentrations required to stimulate retinoic acid receptor-mediated transcription *in vitro*. *J Clin Invest* 90: 1269–1274

52   Chambon P (1995) The molecular and genetic dissection of the retinoid signaling pathways. *Recent Prog Horm Res* 50: 317–332

53   Marcelo CL, Dunham WR (1997) Retinoic acid stimulates essential fatty acid-supplemented human keratinocytes in culture. *J Invest Dermatol* 108: 758–762

54   Punnonen K, Puustinen T, Jansen CT (1988) The antipsoriatic drug metabolite etretin (Ro 10-1670) alters the metabolism of fatty acids in human keratinocytes in culture. *Arch Dermatol Res* 280: 103–107

55   Ponec M, Boonstra J (1987) Effects of retinoids and hydrocortisone on keratinocyte differentiation, epidermal growth factor binding and lipid metabolism. *Dermatologica* 175 (Suppl 1): 67–72

56   Imakado S, Bickenbach JR, Bundman DS, Rothnagel JA, Attar PS, Wang XJ, Walczak VR, Wisniewski S, Pote J, Gordon JS et al (1995) Targeting expression of a dominant-negative retinoic acid receptor mutant in the epidermis of transgenic mice results in loss of barrier function. *Genes Dev* 9: 317–329

57   Attar PS, Wertz PW, McArthur M, Imakado S, Bickenbach JR, Roop DR (1997) Inhibition of retinoid signaling in transgenic mice alters lipid processing and disrupts epidermal barrier function. *Mol Endocrinol* 11: 792–800

58   Yaar M, Stanley JR, Katz SI (1981) Retinoic acid delays the terminal differentiation of keratinocytes in suspension culture. *J Invest Dermatol* 76: 363–366

59   Choi Y, Fuchs E (1990) TGF-beta and retinoic acid: regulators of growth and modifiers of differentiation in human epidermal cells. *Cell Regulation* 1: 791–809

60   Jetten AM (1990) Multi-stage program of differentiation in human epidermal keratinocytes: regulation by retinoids. *J Invest Dermatol* 95: 44s–46s

61   Tong PS, Horowitz NN, Wheeler LA (1990) Trans retinoic acid enhances the growth response of epidermal keratinocytes to epidermal growth factor and transforming growth factor beta. *J Invest Dermatol* 94: 126–131

62   Magnaldo R, Bernerd F, Asselineau D, Darmon M (1992) Expression of loricrin is neg-

atively controlled by retinoic acid in human epidermis reconstructed *in vitro*. *Differentiation* 49: 39–46

63   Varani J, Nickoloff BJ, Dixit VM, Mitra RS, Voorhees J (1989) All-trans retinoic acid stimulates growth of adult human keratinocytes cultured in growth factor-deficient medium, inhibits production of thrombospondin and fibronectin, and reduces adhesion. *J Invest Dermatol* 93: 449–454

64   Sanquer S, Gilchrest BA (1994) Characterization of human cellular retinoic acid-binding proteins-I and -II: ligand binding affinities and distribution in skin. *Arch Biochem Biophys* 311: 86–94

65   Marcelo CL, Duell EA, Rhodes LM, Dunham WR (1992) *J Invest Dermatol* 99: 703–708

66   Marcelo CL, Rhodes LM, Dunham WR (1992) *J Invest Dermatol* 103: 564–568

Index

PIR
Progress in Inflammation Research

Medicinal Fatty Acids in Inflammation

Kremer, J.M.,
Albany Medical College, Albany, USA (Ed.)

This volume is a unique assembly of contributions focusing
on the biochemical, immunological and clinical benefits of
n-3 fatty acids in inflammation.

Leading clinical investigators from fields as diverse as
rheumatology, dermatology, nephrology, gastroenterology
and neurology have authored chapters. The basic scientific
underpinnings of their findings are elucidated as well.

The work is a highly accessible, one-of-a-kind source
which will well serve lipid researchers, graduate students,
dieticians and members of the food industry.

PIR – Progress in Inflammation Research
Kremer, J.M. (Ed.)
Medicinal Fatty Acids in Inflammation
1998. 154 pages. Hardcover
ISBN 3-7643-5854-8

Contents

List of contributors

Preface

Calder, P. C.:
n-3 Polyunsaturated fatty acids and mononuclear
phagocyte function

Zurier, R. B.:
Gammalinolenic acid treatment of rheumatoid arthritis

Ziboh, V. A.:
The role of n-3 fatty acids in psoriasis

Horrobin, D. F.:
n-6 Fatty acids and nervous system diorders

Fernandes, G.:
n-3 Fatty acids on autoimmune disease and apoptosis

Belluzzi, A. and Miglio, F.:
n-3 Fatty acids in the treatment of Crohn's disease

Rodgers, J. B.:
n-3 Fatty acids in the treatment of ulcerative colitis

Geusens, P. P.:
n-3 Fatty acids in the treatment of rheumatoid arthritis

Grande, J. P. and Donadio, J. V.:
n-3 Polyunsaturated fatty acids in the treatment of
patients with IgA nephropathy

Subject index

BioSciences with Birkhäuser

(Prices are subject to change without notice. 10/98)

For orders originating from all over the world
except USA and Canada:

Birkhäuser Verlag AG
P.O. Box 133
CH-4010 Basel / Switzerland
Fax: +41 / 61 / 205 07 92
e-mail: orders@birkhauser.ch

For orders originating in the USA and
Canada:

Birkhäuser Boston, Inc.
333 Meadowland Parkway
USA-Secaucus, NJ 07094-2491
Fax: +1 / 201 348 4033
e-mail: orders@birkhauser.com

Birkhäuser

PIR
Progress in Inflammation Research

Chemokines and Skin

Kownatzki E. / Norgauer J.,
Albert-Ludwigs-Universität, Freiburg, Germany (Ed.)

The present volume summarizes the state of information on chemokines focussing on skin diseases. The first three chapters deal with the structure and molecular biology of chemokines and their receptors. The following three review information on the interaction of chemokines with lymphocytes, mast cells and eosinophilic granulocytes. One chapter deals with the expression of chemokines in several inflammatory skin diseases. The final chapter reports on in vitro evidence for a growth-promoting activity of chemokines in skin-derived tumor cells.

The volume is of use for the basic scientist interested in practical aspects and for the physician in search for basic mechanisms of skin diseases.

Contents

PIR · Progress in Inflammation Research
Kownatzki, E. / Norgauer, J. (Ed.)
Chemokines and Skin
1998. 140 pages. Hardcover
ISBN 3-7643-5818-1

BioSciences with Birkhäuser

(Prices are subject to change without notice. 10/98)

For orders originating from all over the world except USA and Canada:

For orders originating in the USA and Canada:

Birkhäuser Verlag AG
P.O. Box 133
CH-4010 Basel / Switzerland
Fax: +41 / 61 / 205 07 92
e-mail: orders@birkhauser.ch

Birkhäuser Boston, Inc.
333 Meadowland Parkway
USA-Secaucus, NJ 07094-2491
Fax: +1 / 201 348 4033
e-mail: orders@birkhauser.com

Birkhäuser

PIR
Progress in Inflammation Research

Cytokines in Severe Sepsis and Septic Shock

Redl, H. / Schlag, G.†,
Ludwig Boltzmann Institute for Experimental and Clinical
Traumatology, Vienna, Austria

This book deals with the central role of cytokines in the generalized inflammatory response of the host as the consequence of severe infection/endotoxin action. International specialists cover several aspects in 20 chapters starting with the agents responsible (endotoxin, superantigens) and recognition during cytokine induction. Further chapters deal with the signal transduction cascade, its modulation due to sex or genetic polymorphism, and the possibilities and problems in detection (including surrogate markers). Major targets of actions are covered in the chapters on coagulation/fibrinolysis, adherence molecules, vasoactive factors, apoptosis and metabolism. As not all actions of cytokines are beneficial, several chapters deal with the prevention of induction, modulation of the cytokine generation or scavenging cytokines including gene therapy approaches. Models are necessary for obtaining pathophysiological information and for testing therapeutic approaches, and thus all chapters deal with experimental models as well as clinical trials. The reasons why these have failed so far are the subject of the final chapter.

Researchers and students of Critical Care Medicine and Biomedicine will find up-to-date reviews and data in this book.

PIR – Progress in Inflammation Research
Redl, H. / Schlag, G. †
Cytokines in Severe Sepsis and Septic Shock
1998. Approx. 300 pages. Hardcover
ISBN 3-7643-5877-7
Due in November 1998

BioSciences with Birkhäuser

(Prices are subject to change without notice. 10/98)

For orders originating from all over the world except USA and Canada:

Birkhäuser Verlag AG
P.O. Box 133
CH-4010 Basel / Switzerland
Fax: +41 / 61 / 205 07 92
e-mail: orders@birkhauser.ch

For orders originating in the USA and Canada:

Birkhäuser Boston, Inc.
333 Meadowland Parkway
USA-Secaucus, NJ 07094-2491
Fax: +1 / 201 348 4033
e-mail: orders@birkhauser.com

Birkhäuser

PIR
Progress in Inflammation Research

Inducible Enzymes in the Inflammatory Response

Willoughby, D. A., Tomlinson, A.,
Department of Experimental Pathology, The Medical
College of Saint Bartholomew's Hospital, Charterhouse
Square, London, UK (Ed.)

The inducible isoforms of the enzymes cyclooxygenase
(COX 2), nitric oxide synthase (iNOS) and heme oxygenase
1 (HO-1) have generated great interest as possible
therapeutic targets in inflammation. This book is the first
publication to address the importance of all three enzymes
and the consequences of their interactions to the
inflammatory process.
The book brings together overviews by leading researchers
in the field of the current status of knowledge of COX,
NOS and HO in inflammation. These overviews cover a
series of new concepts in the mechanism of inflammation.
Topics include inducible enzyme involvement in
inflammatory processes including the role in vascular
permeability, leukocyte migration, granuloma formation,
angiogenesis, neuroinflammation and algesia. New
findings from transgenic animal models are reviewed.
Other chapters address the importance of these enzymes
in inflammatory disease states including rheumatoid
arthritis, atherosclerosis and multiple sclerosis. The
possibility of selective inhibitors or inducers of COX, NOS
and HO, and their use in the clinic is discussed.
The subject matter of this book is of interest to rheumatol-
ogists, pathologists, pharmacologists, neuroscientists and
anyone with an academic interest in the mechanisms of
inflammation.

Contents

PIR – Progress in Inflammation Research
Willoughby, D. A., Tomlinson, A. (Ed.)
**Inducible Enzymes in the Inflamma-
tory Response**
1998. Approx. 200 pages. Hardcover
ISBN 3-7643-5850-5
Due in November 1998

BioSciences with Birkhäuser

For orders originating from all over the world
except USA and Canada:

For orders originating in the USA and
Canada:

(Prices are subject to change without notice. 10/98)

Birkhäuser Verlag AG
P.O. Box 133
CH-4010 Basel / Switzerland
Fax: +41 / 61 / 205 07 92
e-mail: orders@birkhauser.ch

Birkhäuser Boston, Inc.
333 Meadowland Parkway
USA-Secaucus, NJ 07094-2491
Fax: +1 / 201 348 4033
e-mail: orders@birkhauser.com

Birkhäuser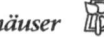